The AI Awakening

A New Era of Data Discovery and Growth

Exploring the Synergy Between Technological Innovation and Human

Dr. Mozhgan Tavakolifard

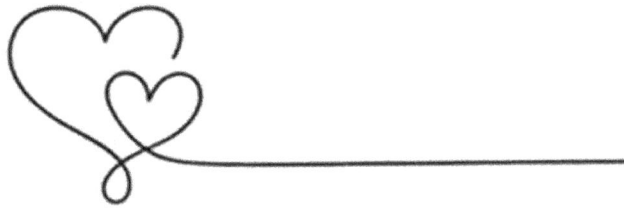

"I dedicate this book to my loving husband, Christian, my wonderful daughter, and my parents. Additionally, I dedicate it to all the fantastic, conscious-aware business leaders who dream of improving the world. Lastly, I dedicate this book to myself, trusting my intuition and following my passion, as I always have."

Foreword

By Professor Letizia Jaccheri

Letizia Jaccheri is a Professor of Software Engineering at the Norwegian University of Science and Technology (NTNU). With a Ph.D. from Politecnico di Torino, she has published over 200 papers and leads initiatives promoting gender balance in computer science through her work with EUGAIN. She is also an ACM Distinguished Speaker and has taught software engineering since 1994. Letizia is passionate about bridging the gap between technology and human values, which aligns closely with the themes explored in this book.

Here is what she has to say about Mozhgan Tavakolifard and her work:

"In the autumn of 2012, I attended the Researcher Grand Prix event at Studendersamfundet in Trondheim. This competition featured PhD students who were challenged to present their research in an engaging and understandable way for the general public. I've always had a deep passion for research, education, and communicating complex ideas in simple terms, which is why I regularly attend such events. One of the many presentations that evening stood out to me in particular—Mozhgan Tavakolifard's.

While most other students relied on props or costumes to symbolize their fields of study, Mozhgan's approach was refreshingly different. She used only her voice, body language, and intellect to convey her research on trust in a digital world—a deeply human concept. Her work explored how trust plays a crucial role in making intuitive decisions in unfamiliar situations, something Artificial Intelligence and technology have yet to replicate. This idea resonated profoundly, as I've always been drawn to the intersection of human elements and technology.

Our professional relationship grew stronger when Mozhgan became a postdoctoral researcher in our department. As one of the few female professors, I've always valued opportunities to mentor and collaborate with young female researchers, and working with Mozhgan was particularly rewarding. We shared a passion for fostering inclusion in our field, organizing seminars and conferences to encourage more women to pursue careers in technology and research. Beyond work, we also formed a close personal bond, often going for runs through the scenic landscapes of Trondheim, where we would discuss everything from our professional lives to personal philosophies.

Mozhgan has worked with various organizations throughout her career, from medium-sized businesses to major Fortune 500 companies. After founding two startups in 5 years, she transitioned into executive roles, serving as Chief Data Officer (CDO) and Chief Analytics Officer (CAO) in the utility, green energy, and retail sectors. Her primary focus has always been leading digital transformations, leveraging data and AI to drive organizational change. However, her approach goes beyond just the technical—Mozhgan excels at creating governance frameworks and building architectures that allow businesses to scale sustainably. She understands that true digital transformation is as much about people as technology.

What sets Mozhgan apart is her human-centered approach. She consistently emphasizes the importance of involving people in every step of the digital transformation process, from strategy development to adopting new technologies. This unique perspective has resonated with businesses of all sizes, reminding them that while technology can enhance operations, the human element ensures lasting success.

More recently, Mozhgan has shifted her focus to helping startups and scale-ups through her company, AI Alchemy Hub, a community designed to nurture a new generation of leaders who aim to create a positive impact. By fostering a space where businesses can grow while contributing meaningfully to society, she is helping shape a future where technology and humanity thrive together.

Her journey from a PhD student exploring trust in a digital world to a leader in data and AI-driven transformation is inspiring. I enjoyed reading Mozhgan's book in September 2024, during a busy time with conferences and research travel. The book is divided into two parts: the first explains the importance of digital transformation, and the second details the steps required to achieve it. What struck me most about her work is how she seamlessly weaves together insights from personal and business transformation, showing how principles of human growth apply to organizational change.

Mozhgan's human-centered philosophy extends beyond her book. Through keynote speeches, podcasts, and Alzen, her latest venture, she has shared her vision of businesses as living ecosystems. This view reflects her belief that personal growth and business evolution are deeply interconnected. Her TEDx talk, "Your Data is Your Currency," further underscores the critical role that personal data plays in the digital economy. Mozhgan's ability to integrate technology with conscious awareness exemplifies her entrepreneurial ventures, such as IPower.me and DharmicData, which focus on sustainable, data-driven growth.

Mozhgan has been recognized as one of Norway's 50 Leading Women in Tech. Her expertise spans industries and regions, from U.S. tech giants to emerging companies. In her book, she provides concrete examples of digital transformation in various sectors, drawing connections between established companies like Google and Uber and newer players such as Jumia in Kenya and Nigeria.

Reading Mozhgan's book has left a lasting impact on me. I feel inspired to share her insights with my students, colleagues, and the companies I work with. It's one of those rare books I wish I had written myself—deeply reflective of the balance between technology, humanity, and growth. The book reinforces my belief in the importance of blending the human and technical aspects of our work and offers practical guidance for anyone involved in digital transformation. I'm excited to pass on her wisdom and continue exploring the interplay between personal development, organizational growth, and technology."

Contents

Preface

A Journey of Transformation

Have you ever felt like you have been on a long journey, searching for perfection, yet it remains out of reach? Avicenna's words at the beginning of this chapter capture this beautifully. This book is our journey through the vast deserts of data and artificial intelligence (AI) as we seek knowledge and a transformative understanding.

Imagine a world where data and AI transform what were once just machines into dynamic, insightful tools. Welcome to a new era where technology meets human potential that will craft a vibrant, innovative future.

I'm Mozhgan Tavakolifard. I have walked this path, from a childhood enchanted by the elegance of mathematics to becoming a pioneer in data-driven innovation. I invite you to join me in exploring how the intricate dance of data and AI can transform our world—and ourselves. What is at stake is how we, as a society, harness the power of data and AI to drive pioneering innovation and shape a future that positively impacts humanity.

When I set out to write this book, I was reminded of the journey that has led me to the intersection of technology, business, and human potential. It is a curious and passionate one that has included a relentless pursuit to understand how data and AI can reshape the fabric of our society. While writing its final stages, I sought the seclusion and inspiration of the coastal town of Kragerø, Norway, retreating to the historic Victoria Hotel—once frequented by Edvard Munch, whose profound influence on the Expressionist movement resonates through masterpieces like *The Scream* and *The Sun*. This setting, steeped in artistic legacy, became a mirror reflecting my creative journey.

Every morning, I woke to *The Sun* above my bed, a replica of Munch's radiant mural, igniting my thoughts with its intense, vibrant energy. Surrounded by landscapes that fueled Munch's artistry, I found myself walking the line between his world and mine, seeing through his eyes—the serene seascapes and vibrant life of a town that echoes the past yet pulses with the present.

Immersing myself in Munch's life and understanding his artistic dedication and personal struggles helped me connect deeply to my work. Much like Munch, I realized that my writing is not just a pursuit of knowledge but a quest to capture the essence of human experience—in the age of data and AI.

This revelation deepened my resolve to infuse my insights with humanistic integrity. It reassured me that patterns and algorithms form a narrative as compelling and intricate as any painted by Munch. This understanding, gained in the quiet reflection of Kragerø, has woven a more decadent, more nuanced layer into the fabric of this book, reinforcing my commitment to exploring the transformative powers of technology with authenticity and emotional depth.

I invite you to join me on this extraordinary journey fueled by curiosity, passion, and an unwavering belief in the power of data and artificial intelligence. From my childhood years, marked by a love for mathematics and a fascination with higher dimensions, to my transformative endeavors in data-driven business growth, I have journeyed as Mozhgan Tavakolifard—an incubator, a dreamer, and an alchemist for change.

Seeds of Curiosity

From an early age, I was captivated by mathematics. However, as I grew, I realized that numbers could tell stories, drive change, and transform industries. My understanding of transformation through data and AI is rooted in my early life. Mathematics became my sanctuary, a realm with a seamless interplay of logic and creativity. As I delved deeper, my love for geometry led me to explore the mysteries of higher dimensions. Armed with nothing but a pen and paper, I embarked on a quest to solve intricate problems in these abstract spaces. In higher-dimension geometry, my visionary talent began to develop and take shape. However, this was the beginning, and my vision evolved and expanded.

Nurturing the Vision

My journey exploring higher dimensions fostered in me a unique visionary perspective. I navigated abstract landscapes, tackling complex problems that stretched the boundaries of my imagination. Hyperspheres (spheres in higher dimensions), polytopes (geometric objects with flat sides), and other intricate structures became subjects of mathematical expeditions. Through rigorous exploration and relentless pursuit, I honed my ability to perceive patterns, connections, and possibilities beyond the constraints of traditional geometry. It was in this realm that my talent flourished. However, as profound as these abstract realms were, another horizon beckoned—merging the abstract with the tangible.

As we navigate the transformative journey of data and AI in business, each section of this book concludes with a practical activity, question, or problem to consider. The purpose of these exercises is to encourage self-reflection and application of the concepts discussed. They will help introspectively assess your experiences and strategies in dealing with challenges and venturing into unknown territories. Engaging in them will not only reinforce understanding but also aid in translating theoretical insights into actionable steps for personal and organizational growth. **Here is the first one:**

Activity:

Have you ever delved deep into a subject only to discover a passion and vision that you never knew existed within you?

The Convergence

The transition from geometry to the world of data and AI was a natural evolution of my thinking. Guided by a belief in the transformative power of these fields, I embarked on a new journey—one of data-driven transformation. Using the analytical skills and intuition I developed exploring higher dimensions, I leveraged the power of data to uncover hidden insights, anticipate emerging trends, and drive innovation. With this newfound knowledge and perspective, I am poised to make a tangible impact in the real world.

Activity:

As you think about the evolving landscape of technology and business, where do you see the convergence of your passions and the opportunities presented by data and AI?

Pioneering Innovation

Armed with a perspective and determination, I embarked on innovating. I became a pioneer, guiding organizations toward a future shaped by data-driven transformation. By perceiving patterns and connections in vast datasets, I helped organizations make informed decisions, optimize operations, and unlock new possibilities. A vision of a world empowered by the untapped potential of data became my driving force. However, I realized that my journey was not just about innovation but also about inspiring others.

Activity:

How do you envision data and AI pioneering innovation in your industry? What transformative changes do you anticipate?

Inspiring the Future

As my journey continues, I remain committed to inspiring the next generation of visionaries. By nurturing curiosity, embracing creativity, and fostering a deep understanding of the power of data and AI, we can shape a future that transcends our current limitations. This journey, filled with challenges and discoveries, inspired me to capture my experiences and insights.

Activity:

Who has been a source of inspiration in your life? How do you envision inspiring others in your field or community?

Why I Wrote This Book

This book provides a comprehensive guide to leveraging data and AI for meaningful organizational transformation. It is not just about theory; it is about practical application. Whether you are a business leader, an aspiring data professional, or simply curious, this book will equip you with the knowledge, tools, and holistic approach needed to thrive in the data-driven era.

It culminates my years of experience, research, and countless conversations with professionals, perhaps like yourself, who seek guidance in navigating the ever-evolving landscape of data and AI. I want to share the unique perspective I have gained from my transformational journey, from a curious child fascinated by the mysteries of higher dimensions to a leader driving data-driven innovation. **What my book offers is:**

- **A holistic approach**: Unlike many other books on data and AI that focus solely on technical matters or theoretical concepts, I take a holistic approach. Successful data and AI initiatives require a multifaceted understanding encompassing strategy, culture, talent, technology, and ethics. Addressing these interconnected elements provides a comprehensive perspective that drives meaningful organizational change.

- **Practical insights**: This book is not just a theoretical exploration of data and AI; it is a helpful guide that empowers you to take action. Its chapters share real-world case studies and actionable tactics that can be applied directly to projects and initiatives. The goal is to bridge the gap between theory and practice, helping you navigate the complexities of implementation and achieve tangible results.

- **Thoughtful reflection**: This book encourages thoughtful reflection in addition to practical insights. It delves into the broader implications of data and AI while exploring ethical considerations, societal impact, and the future of work. Engaging in such critical conversations drives a more nuanced understanding of the opportunities and challenges presented by data and AI.

- **Audience focus**: I aim to meet the needs of a diverse audience. Whether you are a business executive seeking to unlock the value of data, a data professional looking to enhance skills, or a decision-maker evaluating the potential of AI, this book provides valuable insights and guidance. It is written in a clear and accessible manner, ensuring that technical concepts are explained in a way that anyone can understand.

Activity:

How would a holistic approach to data and AI applications benefit your organization or industry? Could a more comprehensive understanding lead to breakthroughs in these areas?

The Book's Structure

Divided into two main parts, the book begins with stories that link data development and AI with personal growth and organizational change, exploring how a growth mindset—viewing challenges as opportunities for development—can catalyze transformation in the digital age.

The initial chapters lay the groundwork for understanding the disruptive nature of data and AI. They explore the essence of awakening to change and cultivating a growth mindset, which is essential for navigating the complexities of today's digital landscape. This framework, born from my experience, guides readers through overcoming obstacles and achieving their full potential. It is enriched with real-world anecdotes highlighting the tangible impact of data and AI.

This book is a call to action for leaders poised to make a positive impact. It offers insights into achieving transformative growth without necessarily facing hardship.

Part One: Transformation: What It Is and Why It Matters

This lays the groundwork for understanding transformation in both personal development and organizational change. Beginning by exploring the awakening to transformation, it progresses through cultivating a growth mindset specific to data-led transformation, harnessing the power of data for insightful decision-making, and exploring the impact of AI. As we delve into disruptive technologies and their role in prompting transformation, I introduce the Wisdom Spire Framework—a structured approach to understanding and navigating the complexities of transformation.

Part Two: The Wisdom Spire Framework: Transformation in Action

The second part practically applies the Wisdom Spire Framework, guiding you through the strategic and operational aspects of implementing data and AI within an organization. Starting with overarching strategies, it moves through tracking value in data and AI initiatives, blueprinting and implementing data-driven technology ecosystems, and establishing a robust data foundation. Furthermore, it discusses building a data-driven organization, navigating effective work methods in AI, and ensuring responsible AI governance. This part discusses harnessing predictive foresight alongside human intuition, bringing a holistic perspective.

The Wisdom Spire Framework, covered entirely in Chapter 6, provides a comprehensive approach to organizational transformation by integrating technical, strategic, and human elements. The framework guides organizations through the complexities of modern business challenges by ensuring that transformation is not just about change but about inspiring a meaningful, sustainable evolution. **The critical components of the Wisdom Spire Framework are:**

- **Foundation (establishing the base camp):** The first step is to build a solid foundation by breaking down data silos. Implement a centralized data platform that integrates information from all departments, enabling a unified view of the company's operations. This is essential for creating a data-driven culture and ensuring that decisions are based on comprehensive insights.

- **Alignment (building the organizational spire):** Restructure the organization to be more agile and responsive to change. This involves flattening the hierarchy, promoting cross-functional teams, and fostering a culture of innovation. Introduce agile methodologies across the organization, which allows for faster iteration and adaptation in response to market changes.

- **Empowerment (elevating through the wisdom spire):** Empowering the workforce is crucial for successful transformation. Provide extensive training on new systems and tools, ensuring employees adapt to changes and actively contribute to the transformation process. Implement a continuous feedback loop in which employee insights are used to refine and improve the transformation strategy.

- **Innovation (reaching the summit):** Finally, focus on innovation. The company can explore new business models and digital products with new systems and structures. Advanced analytics and AI can identify new opportunities to expand the company's market reach significantly.

The Wisdom Spire Framework is practical because it addresses specific organizational challenges during transformation. It emphasizes flexibility, adaptability, and the integration of human expertise with technology, ensuring that the transformation is sustainable and aligned with long-term strategic goals.

By navigating these two parts, you will understand how transformative technologies can be integrated into personal growth and organizational success. Each chapter deepens knowledge and includes practical activities and reflective questions to foster a personal connection to the material, ensuring that the concepts discussed are understood and actionable.

The Journey Ahead

As we set forth together, I invite you to open your mind, embrace curiosity, and prepare for a transformative exploration of data and AI. You will gain the knowledge, tools, and inspiration needed to harness the power of data and artificial intelligence. By leveraging practical insights, holistic approaches, and thought-provoking reflections, you will become equipped to navigate the complexities of implementation, achieve tangible results, and become a catalyst for change within your industry. Welcome to a world of transformation, where possibilities are boundless, and the future is waiting to be shaped.

The potential of data and AI is vast. By joining me on this journey, you are taking the first step towards unlocking that immense potential.

Mozhgan Tavakolifard

PART ONE

Transformation: What It Is and Why It Matters

CHAPTER 1

Awakening to Transformation, Both Personal and Organizational

With over two decades of experience in academia, industry, and entrepreneurship, I can attest to the transformative power of data and AI. My work, which includes leading projects, publishing research, and speaking globally, has not only focused on technological advancements but also personal growth. My experiences have made me realize that transformation extends beyond the digital realm. It is about creating a better future for everyone, one in which technology is crucial to helping us achieve our goals.

This realization marked a pivotal shift in my career and mindset. My role expanded beyond purely technical aspects. I began to grapple with broader questions: How should the team be structured? What competencies do we need? How should we work together? What infrastructure and strategies are necessary? These were not just technical questions—they required a more holistic approach, encompassing leadership, strategy, and organizational dynamics.

As I took on these new challenges, I discovered a passion for making an impact that went beyond just building products or solutions. I enjoyed the creation process of bringing together different elements—people, technology, strategy—to create something meaningful and lasting. This began a journey that would eventually lead me to entrepreneurship.

This chapter explores awakening to transformation and how it applies to personal development and organizational change. It will include the importance of a growth mindset in navigating the complexities of the digital age and drawing parallels between personal enlightenment and business innovation. Inspired by Vishen Lakhiani's work, the chapter outlines four stages of organizational awareness. It covers the importance of challenging old beliefs and social norms to grow personally and as an organization.

Transformation is more than adopting new technologies; it is about integrating our personal and professional selves and fostering a culture of continuous learning, creativity, and innovation.

The Journey of Transformation

Life is a complex web where changing ourselves and our businesses go hand in hand. In reflecting on my personal growth journey, I draw parallels to the transformative journey of companies like Microsoft. Like starting as a small startup and evolving into a mature business that included substantial shifts in business focus, my path has been marked by continuous learning, growth, and evolution.

Microsoft's journey from a modest beginning to a global leader with over 212 billion dollars in revenue and 221,000 employees showcases a remarkable transformation in technology and business. Though different in scale, my transformation mirrors this evolution.

I have had to redefine myself just as Microsoft redefined its business scope and impact. My journey—from focusing on specific tasks and objectives to embracing a broader vision of purpose and effect—reflects a shift in mindset, like Microsoft transcending its initial scope to become a leader in diverse technological fields. This parallel illustrates the transformative power of growth and adaptation, both in personal development and in the business world. Looking back at my experiences, I am amazed at how similar the paths of personal and business change can be.

Figure 1. Awakening Amidst the Ordinary: The Moment of Realization

The transformation journey often begins with a spark, a moment of realization that ignites the path forward. Just as personal journeys, where hardship can ignite a transformative spark, companies like Airbnb (Fisher College of Business n.d.) have also found defining moments in times of economic downturn. This resemblance is not just superficial; it is rooted in a shared experience of facing constraints and challenges—such hardships led to a profound personal and professional transformation, reshaping my approach to life and business. In Airbnb's case, the financial crisis of 2007–2008 served as a catalyst. With limited resources and a challenging market, it turned constraints into a creative force by innovating and pivoting its business model. This resilience and adaptability did not just help them survive; it revolutionized the hospitality industry. Their journey demonstrates the power of embracing challenges as opportunities to innovate and transform.

The awakening I experienced was not a sudden event but a gradual unfolding, a blossoming of understanding that permeated every aspect of my life. It taught me the power of surrender and the importance of love. In my journey, the concept of surrender emerged as a pivotal game-changer. This notion extends beyond personal growth and is crucial in business transformation. I will delve deeper into this idea in future chapters, but for now, let us consider its initial implications. In business, true surrender involves stepping back and allowing data to illuminate the path. It is about shifting from a model where decisions are made solely based on preconceptions and experiences to one where data-driven insights shape choices. This means letting the data reveal what the product or service should be and understanding customer needs and desires not through assumptions but through analysis.

Embracing this approach can be transformative. It encourages businesses to remain agile and responsive, pivoting based on actual, quantifiable customer behaviors and market trends rather than on gut feelings or untested theories. This form of surrender does not imply passivity; rather, it represents a dynamic engagement with data, where insights gained from analytics lead to more informed, effective, and customer-centric decisions. By surrendering to the wisdom of data, businesses can craft products and services that resonate more deeply with their audience, ensuring relevance and sustainability in an ever-evolving marketplace.

Similarly, love's importance in personal transformation parallels the business world. In a corporate context, love can be seen as a deep passion for one's work, a genuine care for the well-being of employees, and a commitment to creating value for customers and the wider community. This form of love fosters a positive corporate culture, drives innovation, and builds solid and authentic relationships with all stakeholders. Businesses that operate with this sense of love and care are often more resilient, enjoy higher levels of employee engagement, and create a loyal customer base. They transform economically and culturally, building a legacy that transcends profits.

In my journey, I encountered a paradoxical truth beautifully illustrated in Jostein Gaarder's book *Sophie's World*. This paradox lies in the realization that we are simultaneously in control and not in control of our lives and circumstances. In the book, Gaarder explores philosophical concepts through the eyes of a young girl, Sophie, who begins to understand the complexities of life and the limits of human understanding. This resonates deeply with my experiences, both personally and professionally.

This paradox manifests in how we make decisions and plan for the future in the business world. We exert control through strategic planning, data analysis, and meticulous execution. However, at the same time, there are myriad factors beyond our control—market shifts, technological advancements, and unforeseen global events like economic downturns or pandemics. This understanding is crucial for business leaders. It teaches us to be prepared, plan, strategize, and stay adaptable, flexible, and open to change. Acknowledging this paradox enables us to navigate the business landscape with a balanced perspective, making informed decisions while remaining agile enough to pivot when the unexpected occurs.

Figure 2. From Constrained Streams to the Ocean of Growth and Transformation

This awakening was challenging. The most significant was learning to be comfortable with not knowing everything and life's inherent uncertainty. In my journey, I learned a vital lesson: stop swimming against the current and instead go with the flow of the river of life. This metaphor for embracing life's unpredictability is intrinsically linked to "being comfortable with not knowing." Both ideas underscore the importance of adaptability and acceptance in the face of life's inherent uncertainties.

In business, as in life, the river's current represents ever-changing market conditions, customer preferences, and technological advancements. Swimming against such currents can be likened to resisting change and clinging to outdated models or strategies that no longer serve in an evolving landscape. On the other hand, going with the flow signifies embracing such changes, adapting strategies, and being open to new possibilities—even when the path ahead is not fully known.

The idea of "being comfortable with not knowing" complements this approach. It is about acknowledging that we cannot predict every turn in the river or foresee every challenge on the horizon. It involves cultivating a mindset prepared for the unknown, equipped to handle surprises, and agile enough to adjust course when necessary. This mindset is critical for businesses navigating the complexities of the digital and data-driven era, where change is the only constant. Such alignment with the flow of life and comfort with uncertainty is critical to thriving in an ever-changing world.

While individual practices like meditation, mindful breathing, journaling, physical exercise, and spending time in nature contribute to personal inner calm and stillness, similar principles can be applied in the business world, particularly in data management and analysis.

Just as mindful practices help individuals see through the noise of daily life to find clarity, businesses can apply similar principles by merging, cleaning, and structuring data from various sources. This process is akin to cultivating corporate mindfulness, where the organization becomes adept at observing and understanding the patterns, trends, and insights that emerge from its data landscape.

For instance, merging data from different departments can provide a holistic view of a business, just as meditation integrates various aspects of our experiences into a unified sense of self. Cleaning and structuring this data is similar to journaling, where organizing thoughts leads to more precise understanding and insights.

Furthermore, as regular physical exercise strengthens the body and improves overall health, consistently analyzing and acting on data insights enhances a business's decision-making processes and strategic health. Akin to the rejuvenating effects of spending time in nature, immersing in the natural flow of data can help companies stay grounded and connected to their market environment, enabling them to adapt swiftly and effectively to changing conditions.

In essence, these data management practices are not just technical tasks; they represent a deeper, more mindful approach to business. By adopting these methods, companies cultivate an awareness that allows them to see through the chaos of an ever-changing market, identify growth opportunities, and make informed decisions that align with their strategic goals.

Throughout this journey, I was fortunate to have the support of mentors and coaches. Their guidance was invaluable, embodying the wisdom of the Shu Ha Ri concept (which I cover more fully later in the book)—that teachers appear when the student is ready and disappear when the student is prepared to move forward independently.

Lakhiani's framework for structuring life goals has been pivotal in my development. He categorizes life goals into three primary facets: growing, experiencing, and contributing. Moreover, he suggests personal key performance indicators (KPIs) across 12 aspects of life. This framework, while individual, offers valuable insights when applied to the business world.

1. **Growing**: In a business context, "growing" mirrors the pursuit of continuous improvement and innovation. Just as personal growth involves expanding our capabilities and understanding, business growth focuses on developing new competencies, exploring emerging markets, and staying ahead of industry trends. Setting KPIs for business growth might include revenue growth, market expansion, or product development metrics.

2. **Experiencing**: Translated to business, this aspect emphasizes creating and delivering enriching experiences for customers and employees. It is about building a brand that resonates emotionally, designing products or services that delight users, and fostering a workplace culture that values and nurtures employees. KPIs could involve customer satisfaction scores, employee engagement, and brand recognition metrics.

3. **Contributing**: In business, contributing takes the form of corporate social responsibility and ethical practices. It is about how a company positively impacts society and the environment. This could involve sustainable practices, community engagement, or ethical supply chains. KPIs could be sustainability goals, social impact initiatives, and ethical compliance standards.

By adopting a framework similar to Lakhiani's—focusing on growing, experiencing, and contributing—businesses can develop a comprehensive set of 360-degree KPIs (Accenture 2021). These KPIs serve as a "North Star" for their strategy, encompassing financial performance and other crucial dimensions like customer satisfaction, employee well-being, innovation, and societal impact. This approach aligns with the modern movement towards holistic value creation in business.

Adopting a 360-degree approach to KPIs allows businesses to align their strategies with broader goals, ensuring that growth is sustainable, offerings are impactful, and contributions are meaningful. This helps achieve long-term success and builds a legacy that transcends financial achievements.

Since my awakening, I have adopted several practices and habits, like organizations adopting best practices to foster growth. My routines include meditation with biofeedback devices and continuous learning. Many forward-thinking companies have recognized the value of allowing employees to dedicate time to personal projects, fostering innovation and constant learning. For example, Google's famous "20 percent time" policy is a prime illustration. This policy encourages employees to spend 20 percent of their work time on personal projects that interest them and could potentially benefit the company. This initiative has created some of Google's most successful products, like Gmail and AdSense.

Another example is 3M's "15% Culture," which permits employees to use 15 percent of their paid time to explore and develop ideas. This policy has been crucial in 3M's consistent innovation output, leading to revolutionary products like Post-it Notes.

Additionally, companies like LinkedIn have implemented "InDays," where employees are encouraged to work on projects—which may not necessarily align with their day-to-day job responsibilities—that they are passionate about once a month. This approach drives innovation and enhances employee satisfaction and retention by acknowledging and nurturing their interests and skills.

By integrating these practices, businesses can create an environment where employees feel valued for their direct contributions to existing projects and their creative potential. This leads to developing new products and services and promoting a workplace culture where continuous learning and innovation are ingrained values.

In my journey, I have encountered the concept of "brules" (i.e., "bullshit rules"). Lakhiani coined the term to describe the societal norms and conventions that often go unchallenged but may not necessarily serve our best interests or true selves. These are the rules that society imposes, which can limit our thinking, creativity, and personal growth. Questioning these brules has been pivotal in finding my path.

Figure 3. Awakening

Lakhiani also introduces the concept of being "unfuckwithable." This term refers to a state of mind where one's sense of well-being and self-worth is not dependent on external approval or societal norms. It is about having such a solid inner grounding and confidence that external events or opinions do not quickly shake you.

One of the most striking examples of brules in the business world is the reliance on gut feelings or unfounded hypotheses about customer behavior rather than data-driven insights. I have frequently encountered this in my work with various businesses. Companies often hold strong opinions about why customers churn, such as beliefs about pricing or product features, without grounding these beliefs in actual data.

For instance, I worked with a company that was convinced their customer churn was primarily due to pricing, which they believed was too high, causing customers to leave for cheaper alternatives. However, the real reasons for churn were uncovered when we conducted an exploratory data analysis. The pricing was not the issue; it was factors like customer service quality and lack of engagement with the product. This revelation was eye-opening for the company, debunking their long-held belief and allowing them to realign their strategies based on factual data.

This example highlights the importance of questioning and challenging such brules in business. Relying on gut feelings without evidence can lead companies astray, causing them to miss out on critical insights that data can provide. By embracing data-driven decision-making and being open to what the data reveals, businesses can overcome these unfounded beliefs and make more informed, effective decisions. This approach is about discarding ineffective habits and adopting a mindset of curiosity and openness to new insights, leading to more robust and successful business strategies. In the chapters ahead, I will further explore these concepts and how they intertwine with the themes I am exploring.

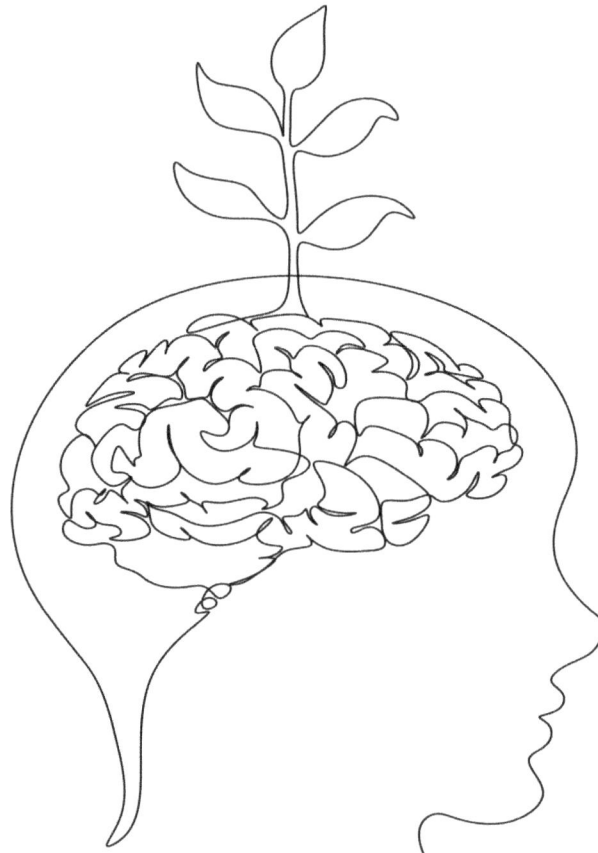

Figure 4. Seeding Innovation, Harvesting Success

To further underscore the importance of a growth mindset in the transformation journey, consider the resilience and adaptability required to navigate the challenges of embracing new technologies and methodologies. This mindset is not just about personal development; it is a crucial component of organizational change, enabling businesses to pivot, innovate, and evolve in response to the dynamic digital environment.

As we delve into the transformation narrative, we must highlight the role of data and AI as catalysts for change. These technologies are not just tools for efficiency; they are the bedrock of innovation, offering insights that challenge our assumptions and push the boundaries of what is possible. Data and AI are the linchpins in this journey, underpinning the strategies that drive personal growth and organizational success.

Cultivating a growth mindset opens the door to leveraging data and AI in ways that transcend traditional applications. It is about fostering a culture in which curiosity drives innovation, challenges are seen as stepping stones to discovery, and integrating technology into strategic thinking becomes second nature. When applied within the context of data and AI, this mindset empowers us to anticipate the future and actively shape it.

Awakening and the Desire for Something More

Every transformational journey, whether personal or organizational, is ignited by a spark—a moment of awakening. For many, this moment arises from unfulfilled longing, a yearning to explore new horizons and seek a deeper purpose.

We cannot create a new future by holding onto the emotions of the past.
—Joe Dispenza (Dispenza 2017)

We cannot move forward if we are stuck in what we used to feel. Big companies have these moments of awakening, too. Even old banks like JPMorgan Chase and Goldman Sachs have had to change (Wosepka n.d.; Kin + Carta 2023; Goldman Sachs n.d.). The advent of banking apps and the transition of many services online are not just mere enhancements in customer service; they represent a profound and fundamental transformation within the legacy banking sector. These developments indicate a more significant shift, which is critical for traditional banks to stay competitive in today's digital-first world.

For instance, mobile banking apps are not just about convenience; they are responses to growing customer expectations for on-demand, accessible financial services. This shift goes beyond just digitizing existing services; it is about reimagining banking in a way that aligns with the digital lifestyle of modern consumers. Banks are moving towards a more integrated digital experience, offering everything from mobile check deposits to real-time notifications and personalized financial insights.

Similarly, bringing more services online reflects the broader digital transformation across all industries. Traditional banks recognize the need to evolve from their brick-and-mortar roots to become more agile, data-driven, and customer-centric. This transition is not just about adopting new technologies; it is about a cultural shift within these institutions, embracing innovation and adapting to their customers' changing behaviors and expectations.

These examples show that legacy banks are not just updating their services; they are undergoing a comprehensive transformation, rethinking their business models, operational processes, and customer engagement strategies. By doing so, they are positioning themselves to stay relevant and competitive in an increasingly digital financial landscape.

Amazon started as just an online bookstore but became the giant it is today by using digital technology well (Palumbo 2021; GlobalData 2023). Netflix has also changed a lot. It went from renting DVDs via snail mail in the United States to streaming shows worldwide (Sharpen n.d.; Alberdi n.d.). These companies exemplify the critical importance of being open to change and new ideas, especially in response to shifts in the market and consumer behavior.

A prime example of this adaptability is Netflix's transformation. Initially a DVD rental service, Netflix pivoted to streaming in response to changing consumer preferences and technological advancements. This shift was partly driven by a decline in subscription numbers for their DVD service, signaling a change in how people consumed media. Recognizing these environmental changes, Netflix embraced innovation and moved towards a streaming model, which has since revolutionized the entertainment industry.

This ability to adapt in response to environmental changes is crucial for long-term success. Businesses that remain rigid in their models and strategies, ignoring market signals and emerging trends, risk obsolescence. On the other hand, companies that are responsive and agile, like Netflix, can capitalize on new opportunities and maintain their relevance in a dynamic market.

Incorporating this perspective emphasizes that openness to change is not merely about embracing new ideas for their own sake but is a strategic response to external environmental shifts. This approach is fundamental for businesses to evolve and thrive in an ever-changing landscape. However, not all companies do well when things change. Sears, which was once a large retail chain, did not do well when shopping moved online (Castus 2021). Likewise, Nokia did not keep up when smartphones came out (Shrivastava & Hawelia 2024; Wang 2022).

For organizations to thrive in this ever-evolving landscape, they must recognize their need to evolve—driven by a desire for growth and innovation. This mirrors my journey, where moments of clarity led to stages of development and enlightenment. Despite the apparent readiness and resources available, very few businesses successfully realize the need for transformation. This rarity of success, confirmed by worldwide surveys and studies, including the McKinsey Global Survey and analysis by Oliver Wyman, highlights a critical issue. The majority fails to transform not due to a lack of budget or technical capabilities but because transformation is often not executed holistically. There is usually an emphasis on innovation and technology but not enough on other aspects of business growth, like new ways of working, altering organizational structure, adopting the right strategy, and change management.

For instance, McKinsey's research reveals that less than one-third of companies report successful transformations that improve and sustain organizational performance (McKinsey & Company n.d.).

Moreover, even successful transformations often fail to fully realize their potential benefits, capturing, on average, only 67 percent of the maximum possible benefits. Oliver Wyman's analysis points out that transformations often falter due to slow momentum, missed targets, and overall fatigue, necessitating a deep dive into a company's DNA to understand and address root causes. Similarly, IMD outlines factors such as lack of a compelling case for change, insufficient focus on co-creation in design, and inadequate attention to culture change as critical reasons for failure (Watkins 2020).

Therefore, a comprehensive and balanced strategy is the key to successful transformation. This involves embracing technological advancements and fundamentally rethinking organizational structures, methods, and cultures. As extensive research highlights, the rarity of successful business transformations underscores a crucial gap in current practices. I aim to bridge this gap by developing a comprehensive framework for transformation centered around data and AI.

This framework, presented throughout the rest of the book, is designed to guide businesses through the multifaceted challenges of transformation, providing a balanced strategy that combines technological innovation with a thorough rethinking of organizational structures, methods, and cultures. By integrating data and AI into the core of this transformation framework, I aim to offer readers a modern, practical approach to achieving sustainable and holistic business transformation.

Activity:

Reflect on a moment in your life or your organization's history that spurred a desire for change. What ignited this awakening?

Four Levels of Awareness for Organizational Transformation

I draw inspiration from Lakhiani's *The Code of the Extraordinary Mind* to map the four levels of awareness regarding the journey of organizational transformation. This mapping is introduced to provide readers with a novel perspective on how organizations can evolve through different stages of awareness, much like individuals do. By applying Lakhiani's principles, which have been transformative personally, to the organizational context, unique insights into how businesses can navigate their transformation journeys can be gained.

This approach will be helpful for readers, as it offers a holistic and human-centric perspective on organizational change. It emphasizes the importance of evolving mindsets alongside business strategies and technologies. Understanding these levels of awareness can equip leaders and decision-makers with a deeper comprehension of the dynamics at play in their organizations, leading to more effective and sustainable transformation.

I propose four levels of awareness. At Level 1, organizations find themselves in their current state, often characterized by data silos and ad hoc analytics tools.

Awakening at Level 2 is when organizations gain access to their data, merge data silos, create and own data platforms or data mesh, and democratize data access. This transition marks the shift from merely having data to unlocking insights and transforming them into valuable information.

Moving to Level 3, organizations connect information and insight to semantic meaning and context through Knowledge Graphs (Google Cloud n.d.-a). This represents the organization's transcendence to the Knowledge level, where deeper understanding and connections are established.

Finally, at Level 4, organizations begin acting based on this knowledge and generate new knowledge. This represents the organization's transcendence to a higher level of wisdom.

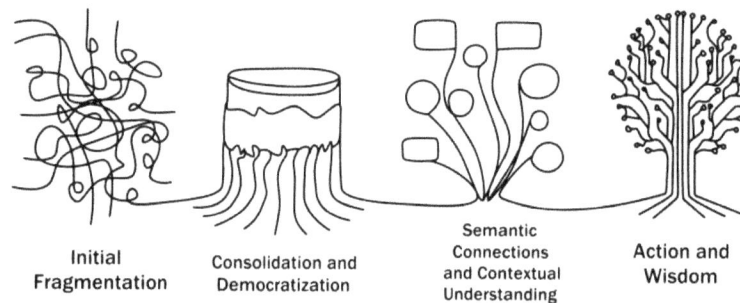

| Initial Fragmentation | Consolidation and Democratization | Semantic Connections and Contextual Understanding | Action and Wisdom |

Figure 5. Data Evolution Journey: From Silos to Wisdom

By understanding and embracing these four levels of awareness, organizations can navigate their transformational journey, evolving from their current state to wisdom and extraordinary impact. Reflecting on these levels, it becomes evident that individuals and organizations are bound by certain beliefs and norms that will either propel them forward or hold them back. Just as individuals seek personal growth and fulfillment, organizations too can embark on a parallel path of transformation. However, to truly transcend boundaries and unlock their true potential, it is essential to challenge and question the very foundations of our beliefs. This brings us to the importance of questioning brules.

In the following sections, while I draw from Lakhiani' 's concepts, my contribution lies in applying and expanding them to organizational transformation. Unlike Lakhiani's primary focus on individual development, I address businesses' unique challenges and opportunities. This adaptation is crucial because it tailors the framework to the specific needs of organizations transforming, particularly in leveraging Data and AI. By doing so, I provide readers with a specialized guide rooted in Lakhiani's philosophy but distinctively focused on business transformation. This perspective is essential for those seeking to apply these principles in a corporate context, especially when driving transformation through data and AI-driven strategies.

Questioning Brules

The first concrete step in personal transformation is questioning and challenging the brules—those limiting beliefs and societal norms we have absorbed. For instance, a typical "brule" is the belief that success is solely defined by wealth and job titles. Another example is the societal norm that dictates that following a traditional career path is the only way to achieve professional fulfillment. By questioning these "rules," individuals can break free from constraints and explore alternative definitions of success and fulfillment, paving the way for more authentic and rewarding personal and professional lives.

- **Origins of brules**: Many of the "rules" we adhere to today have historical or cultural origins, crafted in a different era with different circumstances. Understanding the genesis of these rules can help discern their relevance in the present day. Certain business practices considered standard a few decades ago are now obsolete. For example, manual record-keeping and data analysis were once the norm and have been replaced mainly by advanced data analytics and AI-driven systems. Similarly, traditional marketing methods like print advertising have given way to digital marketing strategies that leverage social media and SEO. Additionally, brick-and-mortar retail models are increasingly complemented by or substituted with e-commerce platforms, reflecting a shift in consumer buying behaviors and the rise of online shopping.

- **The impact of unchallenged brules**: Unchallenged "rules" can stifle innovation, hinder growth, and perpetuate outdated norms. On a personal level, they can limit our potential, keeping us tethered to beliefs that no longer serve our best interests. For organizations, adhering to outdated "rules" can lead to stagnation and prevent them from adapting to the ever-evolving business landscape.

Case Study: Flex Work

Mirvac Group, an Australian property group, challenged the prevailing assumptions about flexible work, which many in the industry believed would decrease productivity and team cohesion. As detailed in an INSEAD case study under CEO Susan Lloyd-Hurwitz, Mirvac found that poor staff engagement and dissatisfaction with work-life balance jeopardized their future success. Only 52 percent of employees felt they could balance work and personal commitments, a figure even lower among construction staff (Kinias & Henderson 2020).

The motivation for introducing the "Smarter Ways of Working" program was multifaceted. Before its implementation, the company participated in a documentary showcasing employees trying flexible work arrangements, providing practical models, and beginning the process of mainstreaming the concept. This initiative aligned with Lloyd-Hurwitz's vision of balancing change management with pushing limits beyond traditional assumptions. It emphasized employee empowerment and introduced new dimensions to flexible work without discarding traditional concepts like remote work.

Moreover, Lloyd-Hurwitz leveraged Mirvac's commitment to sustainability, extending this respect to its workforce. The company believed treating employees with understanding and appreciation was essential for maintaining their best selves and ensuring long-term success.

This case study highlights that the motivation for MMirvac's transition was not just cost-saving or employee retention but also a more profound commitment to employee well-being and organizational sustainability. It demonstrates how questioning traditional norms can lead to innovative solutions that benefit the business and its employees.

Activity:

Challenging Your Brules Identify one "rule" limiting belief you held personally or within your organization. Discuss its origin and impact. Now, brainstorm ways to challenge or overcome this rule. As you reflect, **consider the following guiding questions:**

- What is the source of this rule?

- How has it influenced your decisions or strategies?

- What might be the benefits of challenging this rule?

- Are there data or experiences that contradict this rule?

The Power of a Growth Mindset in the Data and AI Landscape

A growth mindset is rooted in the belief that our abilities can evolve through dedication and continuous learning. It reframes challenges and setbacks as opportunities for growth. This mindset empowers entities to embrace change, innovate, and foster a forward-thinking culture in data and AI. Just as this mindset has pillars supporting its foundation, so does its application in real-world scenarios.

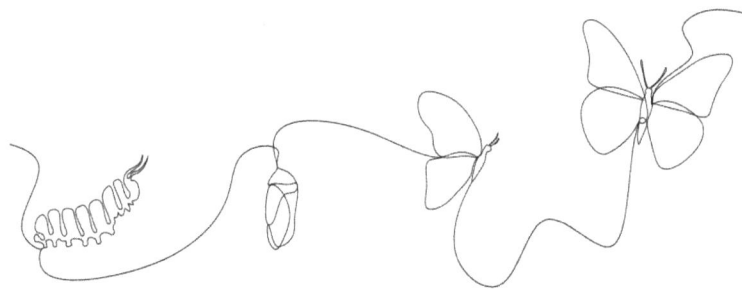

Figure 6. Growth and Transformation Unfolds

Pillars of a Growth Mindset and Implementation

Embrace a Learning Mindset

- **Definition**: Adopt a continuous learning approach. Stay curious, ask questions, and remain updated with the latest trends and advancements in data analysis, AI algorithms, and emerging technologies.

- **In action**: As an entrepreneur, I implicitly grasped this essence. Launching two startups taught me the importance of resilience, the power of embracing failure, and the necessity of continuous learning and improvement.

Embrace Challenges and Learn from Setbacks

- **Definition**: View challenges as growth opportunities. Foster a culture that supports risk-taking, learning from failures, and iterating on ideas.

- **In action**: My entrepreneurial journey with my e-commerce platform is a testament to analyzing gaps, designing new AI models, and overcoming challenges.

Foster Collaboration and Feedback

- **Definition**: Encourage knowledge sharing and value feedback. Create an environment where constructive criticism is seen as an opportunity for growth and improvement.

- **In action**: This pillar was illustrated by collaborating with a data science team to design a new AI model for an e-commerce platform.

Emphasize Agility and Adaptability

- **Definition**: Flexibility and openness to change are crucial. Embrace the iterative nature of data-led transformation and the need to adapt and evolve continuously.

- **In action**: The continuous evolution of companies like Netflix and Amazon embody the principle of adapting to changing landscapes and harnessing data for personalized recommendations.

Championing Innovation

- **Definition**: Cultivate a culture that celebrates innovation. Encourage teams to think outside the box, explore new methodologies, and propose novel solutions.

- **In action**: With its AI-driven Smart Pricing tool, Airbnb showcases the power of championing innovation.

Empowering Employees to Take Risks

- **Definition**: Allow employees the freedom to take calculated risks. Recognize that failure is a part of the innovation process and provide a safety net for them to learn and grow from their experiences.

- **In action**: The creation of Spotify's Discover Weekly feature was a risk that paid off immensely and is a testament to this pillar.

While these pillars provide a theoretical understanding, the Shu Ha Ri philosophy offers a structured approach to mastering them, especially in data and AI. Originating from Aikido, a Japanese form of martial arts, it is a conceptual path toward mastery. In the broadest terms, Shu represents learning and valuing a tradition, Ha challenging and transforming it, and Ri transcending it (Uchitani n.d.).

Conclusion

This chapter has traversed the transformation landscape, guided by the beacon of a growth mindset and illuminated by the transformative power of data and AI. The journey began with an awakening—an acknowledgment of the need for change, both personally and organizationally. Along the path, resilience, adaptability, and a willingness to embrace challenges form the bedrock of true transformation.

By integrating personal and professional selves, the potential to infuse work with passion, purpose, and meaning is discovered. The journey of transformation is inherently intertwined with the journey of growth. Data and AI are not merely tools in this journey but partners guiding us toward insights, innovation, and a deeper understanding of the world.

Moving forward, insights from this chapter will serve as the foundation for embracing continuous learning, creativity, and the pursuit of excellence. Cultivating a growth mindset, paired with the strategic application of data and AI, offers a roadmap for navigating the complexities of our digital era, empowering us to adapt to change and lead it.

Personal, organizational, or technological transformation is a mosaic of challenges, insights, and breakthroughs. It requires continual questioning, learning, and growth. I will now delve deeper into practical applications, exploring how businesses and leaders can harness the principles of a growth mindset and leverage data and AI to navigate transformation successfully. Real-world case studies and actionable strategies will underscore this.

These concepts are not merely theoretical but have practical applications across various industries. For instance, the growth mindset can be seen in how companies like Netflix and Amazon continuously evolve by learning from their data and adopting strategies that foster innovation and adaptation. Similarly, the principles of data strategy and AI transformation can be applied by businesses to optimize operations, enhance customer experience, and drive innovation through predictive analytics, machine learning, and AI-driven decision-making models.

Activity:

Wrapping up this chapter, take a moment to reflect on your personal and organizational journey of transformation. What are the key takeaways for you? How do you envision applying these insights in the coming days?

CHAPTER 2

Cultivating a Growth Mindset for Data-Led Transformation

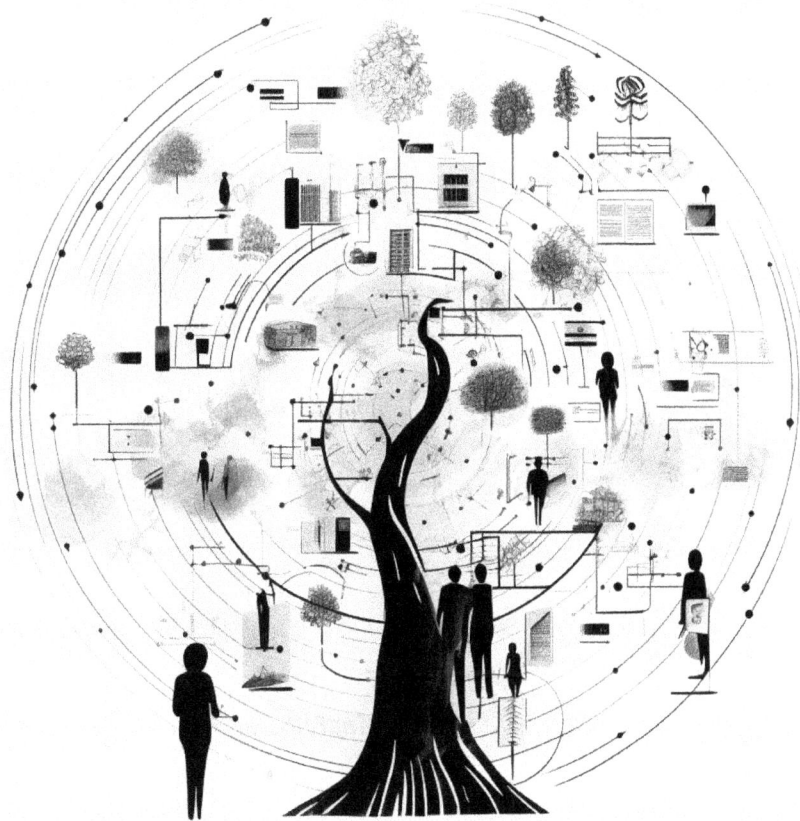

Building upon the intersection of personal transformation and technological advancement, this chapter will explore how technological innovations become the engines of change for individuals and organizations. By delving into the nuances of data's power and AI's potential, their critical role in driving innovation, solving complex challenges, and shaping the future of businesses and societies will be underscored.

When applied to data and AI, a growth mindset enables us to navigate and lead through the complexities of the digital era. It is more than mere adaptation; instead, it is a reimagining of possibilities and redefining of success in an interconnected world.

It is essential to approach the journey with a structured framework for learning and mastery. This brings us back to the concept of Shu Ha Ri—a philosophy that can guide the acquisition of foundational knowledge to the pinnacle of innovation.

Shu Ha Ri: A Framework for Mastery in the Data Realm

Drawing inspiration from Japanese philosophy, Shu Ha Ri provides a framework for understanding the stages of learning and mastery. The Shu stage is the foundation, where individuals and organizations immerse themselves in foundational knowledge. As they progress to the Ha stage, the focus shifts to experimentation. Upon reaching the Ri stage, they are ready to innovate and transcend traditional boundaries.

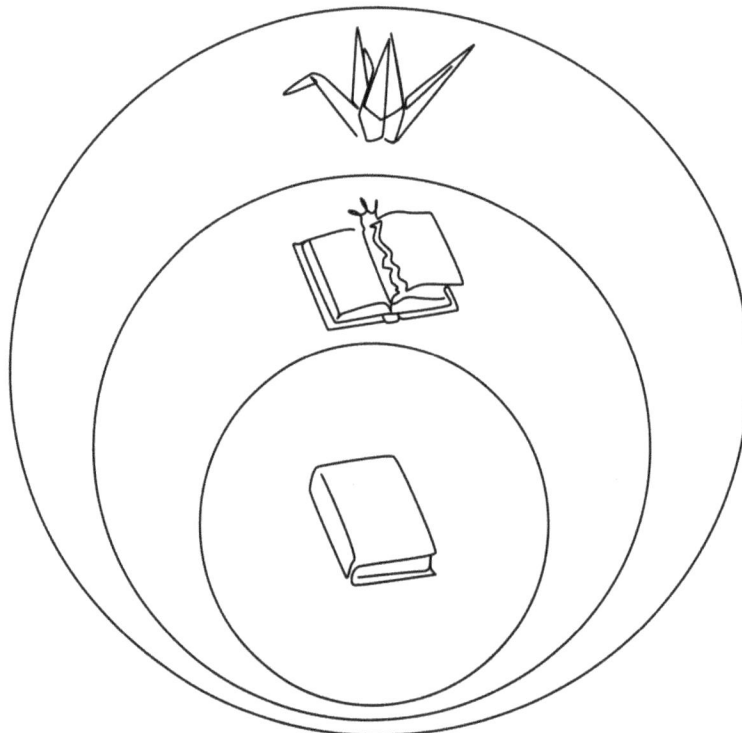

Figure 7. Stages of Mastery: Shu Ha Ri

Shu: The Basics

Learning the basics by following the teachings of experts and masters in the field marks the Shu stage. The goal is to become immersed in foundational knowledge, practices, and techniques related to data analysis, AI, and emerging technologies. Mastering the basics builds a strong foundation for further growth and development.

Ha: Start Experimenting

Once the basics are learned, the Ha stage encourages individuals and organizations to experiment and apply their knowledge in practical contexts. This stage involves branching out, seeking new perspectives, and integrating learning from other experts and disciplines. In data-led transformation, this means exploring different data sources, experimenting with AI algorithms and models, and leveraging insights to drive decision-making and improve processes. It is through experimentation that new possibilities and innovative solutions emerge.

Ri: Achieving Innovation

The Ri stage represents the pinnacle of learning and transformation, where individuals and organizations have transcended traditional boundaries and gained the ability to apply knowledge and skills to various situations. At this stage, imitation has moved beyond, and unique approaches and innovations have been established. Data-led transformation means harnessing the power of data, AI, and advanced analytics to drive innovation, identify new business opportunities, and create value for customers. It is through this stage that transformative change and competitive advantage are achieved.

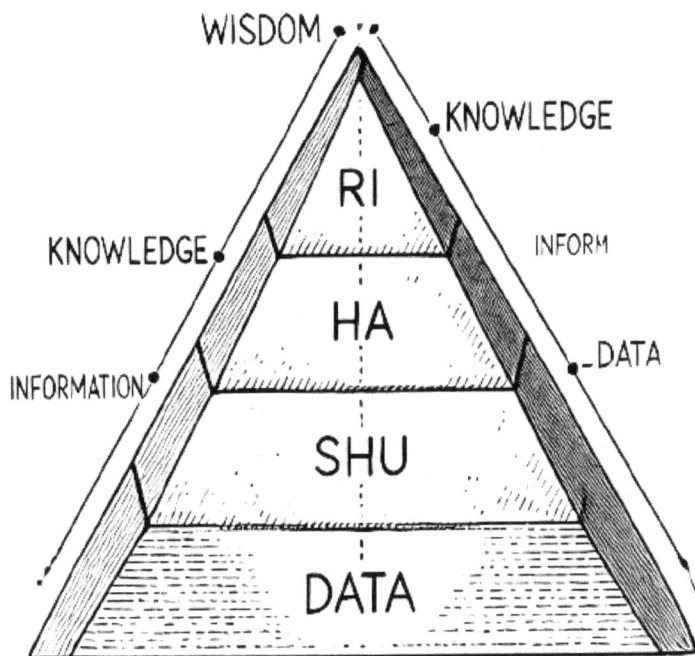

Figure 8. Data-Led Transformation with Shu Ha Ri

This framework has led industry leaders to monumental successes when applied in the corporate world.

Growth Mindset in Action: Industry Leaders' Approach

While the application of data and AI is explored in detail in subsequent chapters, it is worth quickly highlighting how industry leaders have embodied the principles of a growth mindset. These companies have continuously evolved, learned from data, and adopted strategies that can be tied back to the principles of a growth mindset.

Once a DVD rental service, Netflix transformed into a global streaming giant by continuously adapting and harnessing data for personalized recommendations. Similarly, Amazon's journey from an online bookstore to a global e-commerce leader showcases its growth mindset, with AI playing a pivotal role in product recommendations and inventory management.

27

To truly harness this mindset in the context of data and AI, certain elements must be cultivated:

- **Embrace a learning mindset**: Adopt a continuous approach. Stay curious, ask questions, and remain updated with the latest trends and advancements in data analysis, AI algorithms, and emerging technologies.

- **Embrace challenges and learn from setbacks**: View dilemmas as growth opportunities. Foster a culture that supports risk-taking, learning from failures, and iterating on ideas.

- **Foster collaboration and feedback**: Encourage knowledge sharing and value input. Create an environment where constructive criticism is understood as an opportunity for growth and improvement.

- **Emphasize agility and adaptability**: Flexibility and openness to change are crucial. Embrace the iterative nature of data-led transformation and the need to adapt and evolve continuously.

One of the pivotal moments of my career was creating my first startup, Dharmic Data. The vision was to create a personalization platform, primarily as a B2B business aimed at media companies and other industries with large audiences. The goal was to help these businesses gather data, better understand their audiences, and monetize their attention.

Initially, we adopted a hybrid business model that combined consulting with product development. This approach worked well initially, providing us with a steady income stream from consulting, which was critical in funding our product development efforts. However, after two years, it became clear that this mixed model was holding us back.

While lucrative, the consulting side of the business demanded more of our time and resources. We frequently abandoned our product roadmap to fulfill new consulting requests, which often required end-to-end solutions—from data-driven strategy development to use case implementation. While our platform was an accelerator, the consulting work involved much more than product development. This constant shift in focus made achieving the progress we needed on our core product difficult.

Compounding this challenge was our team composition. I had initially hired product experts with a deep passion and skill set for product development. However, they needed to be equipped or motivated to excel in consulting roles. Although we blended different competencies within the team to fulfill consulting tasks, it was clear that this was different from where their true passion lay.

I faced a difficult decision as it became increasingly evident that we needed to split the business so that each side—consulting and product development—could thrive independently. This was not just a strategic pivot but a necessary step to ensure the long-term success of both aspects of our business.

Practical Application in Organizations

To apply the concepts of growth mindset, data strategy, and AI transformation effectively, **organizations should focus on the following steps:**

- **Cultivating a growth mindset**: Encourage continuous learning and adaptability within the organization. Leaders should model a willingness to take risks, learn from failures, and embrace change. This mindset should be instilled at every level of the organization to create an open environment for innovation and new ideas.

- **Developing a robust data strategy**: Start by assessing the organization's current data collection, storage, and usage state. Identify key metrics that align with business goals and ensure that data is accessible and actionable. The organization should focus on breaking down data silos and integrating data across departments to create a unified view to drive decision-making.

- **Driving AI transformation**: Implement AI solutions that complement human expertise rather than replace it. Begin with pilot projects that address specific business challenges and gradually scale AI initiatives. Involve stakeholders from various departments to ensure that AI systems are aligned with business needs and designed to augment human decision-making.

By integrating these concepts, organizations can foster a culture of innovation, leverage data to drive strategic decisions and implement AI technologies that enhance overall business performance.

My Growth Mindset Odyssey: Harnessing Data and AI

My journey exemplifies the transformative power of the growth mindset, especially when combined with data and AI. One project stands out. After launching my second startup, an e-commerce platform, I noticed that our product recommendation system needed to be more effective. Customers were often presented with irrelevant product suggestions, which led to missed sales opportunities and less-than-optimal user experiences.

- **Identifying the challenge**: During the early days of my e-commerce startup, we faced a significant challenge with our product recommendation system. Despite having a wealth of data, our recommendations were often off the mark, leading to missed sales opportunities. I vividly remember one instance where a long-time customer called our support team, frustrated that the system kept suggesting products they had no interest in. This was a turning point; I realized that our data collection was too narrow, focusing only on clicks and purchase history while ignoring crucial signals like browsing patterns and abandoned cart items.

- **Learning and adaptation**: This experience taught me the importance of looking beyond the obvious data points and trusting my intuition when something did not feel right. It also led to a deep dive into

29

our data processes, where we eventually overhauled the system to incorporate more nuanced metrics that better captured customer intent.

- **Analyzing the gaps**: Diving deep into our data collection process, I realized that the system relied heavily on basic metrics like user clicks and purchase history while failing to consider other valuable data points, such as browsing time on products, search queries, and even abandoned cart items. These overlooked metrics could provide a more holistic view of our customers' preferences and behaviors.

- **Designing the new AI model**: Armed with these insights, I collaborated with our data science team to create a new AI model. We incorporated deep learning techniques, which allowed the system to recognize intricate patterns in user behavior. By feeding the model richer data, including the newly identified metrics, it became more adept at understanding user preferences.

- **Performance metrics**: After implementation, we observed a significant improvement in our recommendation system. The accuracy of product recommendations increased by 35 percent, leading to a 20 percent uptick in sales from recommended products alone. Additionally, user engagement with the platform grew, as evidenced by a 15 percent increase in average session duration.

- **Challenges faced**: This journey was challenging. One significant challenge was ensuring data quality. With the inclusion of more data points, there were instances of missing or inconsistent data. We addressed this by implementing a robust cleaning and preprocessing pipeline. Another challenge was the computational demands of the new deep learning model, which was overcome by optimizing the model architecture and leveraging cloud-based solutions for scalable computing power.

This experience vividly reminds me of the transformative power of data and AI. It underscored the importance of continuous learning, adaptation, and the relentless pursuit of improvement—all hallmarks of a growth mindset. A structured approach is paramount to harnessing the full potential of data and AI.

Stepping Into the Unknown

Challenging brules—those societal and self-imposed limiting beliefs—can lead to significant personal and business breakthroughs. For example, challenging a brule in a business context, where the most complex AI solutions are always necessary to solve big problems, can lead to innovative and efficient outcomes.

In the field of data science and AI, there is a prevalent belief that more complex and advanced algorithms yield better results. This belief often drives practitioners to develop intricate and sophisticated models. However, this approach can sometimes overlook advancements in cloud computing, which offer potent resources that can be leveraged effectively.

In one of my professional experiences, I was tasked with addressing a complex issue of workforce optimization for a hotel chain. Traditionally, the approach would have been to develop a single, intricate AI model to tackle the problem comprehensively. However, I decided to take a different path. Instead of creating one complex model, I developed hundreds of simpler AI models, each designed to address specific aspects of the workforce optimization challenge.

Leveraging the power of cloud computing, we ran these models in parallel, utilizing widely accessible resources to achieve a solution that could rival a single, large-scale model. This approach solved the problem effectively and facilitated widespread adoption within the organization.

It demonstrated that the problem may be complex, but the solution does not need to be overly complicated. Sometimes, more straightforward and accessible AI solutions can generate significant value by addressing the core issue in a practical, user-friendly manner.

Figure 9. Moving Beyond Limiting Beliefs

However, what happens when brules are challenged? It often leads to uncharted territory. Stepping into the unknown is scary. It brings up fear of failure and immediately puts us in fight or flight mode, deeply wired into our DNA, a relic of evolutionary history. In the Stone Age, failure had dire consequences, such as being left behind, starving, or facing mortal danger from predators. This ancient context forged a profound aversion to failure, a survival mechanism still active in our psyche.

Such fear manifests differently today, particularly during personal or organizational transformation. Literal survival is no longer at stake, but stress triggers a robust emotional response to setbacks.

As I faced fears and uncertainties during my transformation, people encountered similar challenges when navigating organizational transformations. Recognizing fear as an inherent evolutionary residue can empower confronting and moving beyond it while understanding that significant growth and transformation can occur in unknown and challenging moments.

Organizations often grapple with similar fears regardless of size or industry. Conducting thorough market research can be instrumental in such situations. It helps understand new markets, customer preferences, and potential challenges while reducing uncertainty. Good research equips organizations with valuable insights, aiding in informed decision-making and strategy formulation. However, while market research can mitigate some risks, it cannot eliminate the inherent uncertainties of venturing into new areas. Businesses must balance insights gained from research with a willingness to embrace the unknown and adapt as more is learned.

For instance, startups like Beyond Meat and Impossible Foods, pioneers in substituting plant-based products for meat, boldly entered the competitive food industry. Despite initial doubts, their innovation willingness carved out market share and prompted traditional companies to reconsider their product lines (Torrella 2023; Barkho 2023; Swann & Kelly 2023; Bulah et al. 2023).

Tom Chi, the former product experience visionary at Google and an innovation leader, **encapsulates the spirit of embracing uncertainty and rapid experimentation in the face of the unknown:**

Doing is the best kind of thinking. Maximize the rate of learning by minimizing the time to try ideas. Do not think of theory until it works ten times in practice in several different contexts.
-(TEDx Talks 2014)

This is the spirit demonstrated by companies that successfully pioneer new markets. Such organizations did not shy away from uncharted territory. Instead, they adopted a mindset similar to Chi's: quick prototyping, learning from failures, and iterating swiftly. This approach allows companies to navigate new markets more effectively, turning uncertainty into opportunity and fear of the unknown into a driving force for innovation and market leadership.

Embracing a learning mindset through experimentation and practice is essential in the transformation journey. It requires challenging existing theories, letting go of the fear of failure, and being open to the insights and lessons that emerge from real-world experiences. By doing so, the unknown can be navigated with curiosity, resilience, and a willingness to adapt and grow.

My journey was marked by moments of awakening and transformation, just as the business world is now undergoing a profound shift. Today's tools and technologies are not just about automation or efficiency; they are catalysts for change, reshaping industries and redefining what is possible. My exploration into the depths of self and spirit found a parallel in the vast, uncharted realms of data and artificial intelligence, technologies that—much like the human spirit—can liberate and elevate us to new heights.

My journey took a path of personal growth and professional awakening. It began with a deep curiosity about the patterns and systems that govern our world, which led me to explore various fields and disciplines. This brought me to the frontiers of technology, where I discovered the transformative power of data and AI. I saw how these tools could do more than just process information; they could unlock insights, drive innovation, and catalyze change in unprecedented ways. This realization reshaped my understanding of technology as a tool for efficiency and a catalyst for comprehensive transformation in both personal and organizational contexts.

This parallels a spiritual quest for knowledge and enlightenment that involves navigating complex challenges, encountering moments of doubt, and experiencing profound discoveries. Just as spiritual practices guide individuals towards deeper insights and self-awareness, the data and AI journey leads businesses through a transformative path of better understanding operations, customers, and markets. Ultimately, **clarity and actionable insights for strategic decisions are achieved. As Randy Bean writes:**

> Companies like Amazon, Google, eBay, Facebook, Uber, and Airbnb are rooted in data and analytics and have leveraged new data-driven business models to disrupt and transform traditional industries such as retailing, media, and travel. For innovative firms such as these, data brings speed, agility, and the ability to fail fast, learn from experience, and execute smarter.
> **-(Bean 2021)**

As a spiritual seeker navigates intricate terrain for deeper understanding, businesses traverse complex data landscapes to find insights and clarity. My experiences in navigating these realms, from initial uncertainties to eventual breakthroughs, mirror businesses' journey in harnessing the power of data and AI, marking a transformative quest toward operational and strategic enlightenment.

Activity:

Can you identify any recent challenges or unknown territory you or your organization ventured into? What were the fears or hesitations? How did you overcome them? I think reflecting on this can provide valuable insights for future endeavors.

The Expansion of Data and AI

Industries worldwide are witnessing the transformative impact of data and AI, evident in innovative companies' strategies across different continents. In Africa, the e-commerce leader Jumia has tackled significant logistical hurdles to revolutionize online retail.

One of the critical challenges it faced was the need for proper national address systems in most African countries, coupled with poor road networks. To navigate this, Jumia had to rely on descriptive addresses and landmarks provided by customers, with their delivery personnel maintaining constant contact with clients for directions en route. In markets like Kenya and Nigeria, Jumia adapted the use of motorcycles (*bodabodas*) for swift product delivery within congested cities. Moreover, Jumia implemented a sophisticated system using machine learning (ML) and GPS-enabled delivery apps, with coordinates collected during the first delivery to establish a logistics network for future deliveries (Wawira 2019).

Likewise, digital payment systems like WeChat Pay and Alipay have transformed cash-centric economies in Asia. These platforms, owned by Tencent and Alibaba, have revolutionized the Chinese payment industry by introducing an alternative payment ecosystem that utilizes QR codes and in-app transaction features. This has led to a significant shift from traditional payment processors like card machines and cash registers to smartphone-based transactions, creating new interactions between merchants, consumers, payment providers, and banks. The growth and dominance of WeChat Pay and Alipay stem from integrating existing popular platforms and offering cost advantages to merchants and consumers. The user bases of both platforms have consistently grown, contributing significantly to China's GDP through payment technologies (Caldwell & Liu 2021).

These examples illustrate how leveraging data and AI and adapting to local challenges and consumer behaviors have enabled businesses like Jumia, WeChat Pay, and Alipay to pioneer and lead in their respective markets. They demonstrate the importance of innovative approaches in overcoming logistical challenges and transforming traditional industries in the digital age.

I have been drawn to the transformative potential of data and AI within the business world, discovering how these technologies are not just broad concepts but specific tools for problem-solving and innovation. For example, I observed the use of AI in predictive analytics to forecast market trends and the application of ML algorithms in customer behavior analysis to tailor marketing strategies. My role eventually evolved into that of a guide, helping businesses harness these specific aspects of data and AI to navigate their capabilities better, drive growth, optimize operations, and open new avenues for innovation.

Activity:

List three areas in your personal life or organization where data and AI could bring about transformative change. How would this change look? Could you talk with peers or colleagues to gain diverse perspectives?

Case Study: Data and AI Transformation

I will present a detailed case study of a company I have worked closely with, which I will refer to as RetailCo for confidentiality. Their story demonstrates how a significant shift to a data-driven and AI-enhanced approach can revitalize a stagnant business, transform a Fortune 500 company's traditional business model, and highlight the pivotal role of changing organizational mindsets with technological integration.

RetailCo had faced stagnating growth and recognized the need for a transformative approach. As their consultant, I guided them through a strategic overhaul, focusing on implementing a customer-centric model supported by data and AI. We integrated distributed data architecture to democratize data access across the company and employed solutions such as generative AI (GenAI) to make complex data insights accessible for strategic decision-making. A vital aspect of this transformation was fostering a culture of innovation and entrepreneurial thinking within their data practices.

- **Customer segmentation and personalization**: The company had a wealth of customer data but needed help to leverage it effectively. We began using AI-driven customer segmentation to categorize customers based on purchasing behavior, preferences, and demographics. This segmentation allowed RetailCo to tailor marketing campaigns and personalize customer interactions. For instance, personalized email campaigns based on customer preferences saw a 30 percent increase in open rates and a 25 percent increase in conversion rates.

- **Inventory management optimization**: RetailCo faced challenges with overstocking and stockouts, which impacted its bottom line. By implementing AI-powered demand forecasting models, we analyzed historical sales data, seasonal trends, and promotional impacts to predict future product demand accurately. This optimization reduced inventory holding costs by 20 percent and improved stock availability, leading to a 15 percent increase in sales during peak seasons.

- **Enhanced customer experience with chatbots**: We deployed AI chatbots on RetailCo's website and mobile app to improve customer service. These bots used natural language processing (NLP) to handle common customer inquiries, process returns, and provide product recommendations. This initiative reduced customer service response time by 50 percent and increased customer satisfaction scores by 20 percent.

- **Real-time analytics and reporting**: Management needed timely insights to make strategic decisions. We implemented a real-time analytics dashboard using business intelligence tools integrated with their distributed data architecture. This dashboard provided real-time KPIs such as sales performance, customer engagement metrics, and supply chain efficiency. This visibility allowed RetailCo's executives to identify and address issues promptly, resulting in a 10 percent increase in operational efficiency.

- **Cultivating an innovative culture**: A significant part of the transformation involved shifting the organizational culture. We introduced innovation workshops and hackathons, encouraging employees to develop new ideas and experiment with data-driven projects. One successful project from these initiatives was an AI-powered recommendation engine for their e-commerce platform, boosting online sales by 18 percent.

This comprehensive approach led to significant improvements in customer retention and the acquisition of new customer segments, ultimately resulting in marked business growth. RetailCo's transformation journey is an illustrative example of how integrating data and AI, coupled with a cultural shift, can effectively reignite business growth in today's digital landscape.

Transformations On the Horizon

Continuous personal and organizational transformation is critical when navigating an era characterized by rapid technological advancements, global interconnections, and unprecedented challenges. These evolving landscapes, already upon us, underscore the need for adaptability and forward-thinking strategies to thrive in an ever-changing world.

The ongoing integration of quantum computing, augmented reality, and sophisticated biotechnologies is revolutionizing industries, reshaping economies, and redefining lifestyles. As we continue to navigate this transformative era, it is vital to maintain an open mind and insatiable curiosity.

Tom Chi summarizes the importance of embracing and leveraging these ongoing changes to propel further innovation and growth: "Knowing is the enemy of learning" (Ribeiro 2015).

On a personal level, several critical skills and mindsets are essential for thriving in a world reshaped by automation and AI. These include adaptability (the ability to adjust to new conditions), resilience (the capacity to recover quickly from difficulties), and continuous learning (a commitment to update and expand one's skill set constantly). These attributes and creative problem-solving, emotional intelligence, and collaboration skills will be paramount. The ability to cultivate and leverage these skills and mindsets will determine one's place and success in this rapidly evolving world.

Moreover, global challenges such as climate change, socioeconomic disparities, and geopolitical tensions underscore the need for transformative leaders who can envision sustainable solutions, bridge divides, and inspire collective action. The transformational journeys embarked upon today will shape the leaders of tomorrow.

Understanding and embracing fundamental principles of transformation—such as accepting change, fostering innovation, prioritizing continuous learning, and developing agility—is how to prepare for the immediate future and a world in constant flux. The journey, therefore, is about equipping ourselves and our organizations with the tools and mindsets necessary to thrive in an ever-evolving landscape to ensure long-term sustainability and relevance.

Embracing Wholeness

However, what happens when these two journeys—the personal and the organizational—intersect? As my parallel paths of individual and organizational transformation converged, I realized the importance of embracing wholeness, an integration of my personal and professional self. They ceased to be separate entities and became interconnected aspects of who I am. This integration brought a newfound sense of authenticity, which allowed me to infuse passion, purpose, and deep meaning into my work.

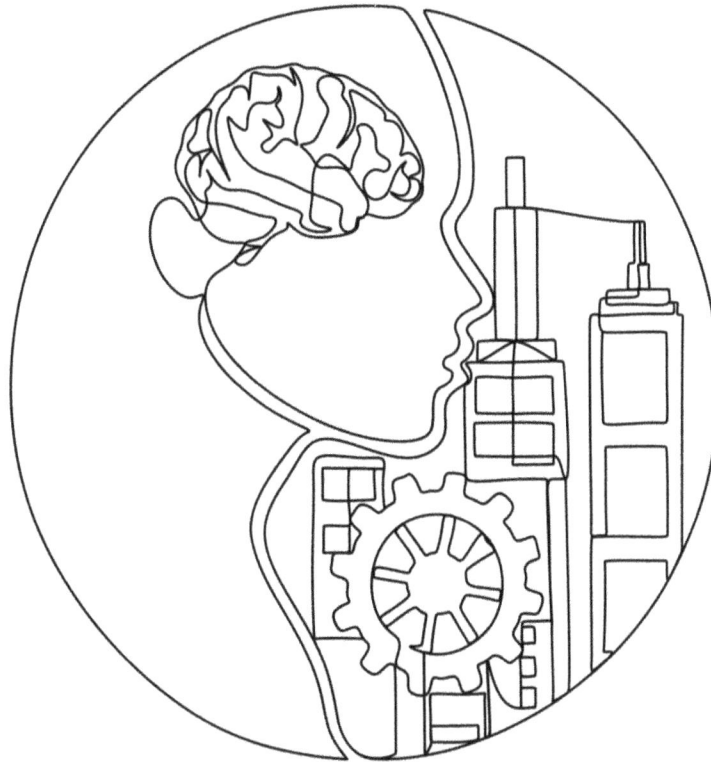

Figure 10. Mind, Mechanism, and Metropolis Merge

Creating Ripple Effects

Just as a pebble creates ripples in the water, my transformation began to radiate outward, influencing those in my immediate surroundings. The insights and lessons of my journey became a beacon for colleagues, clients, and others. This led to a humbling realization: while seemingly individual, personal growth can positively affect our immediate environment in meaningful ways. Reflecting on this, I see how organizational concepts and personal experiences are interconnected and reinforce how personal or organizational growth follows similar patterns.

Individual and corporate transformations are intricately linked—both thrive on growth, discovery and a relentless quest for excellence. The repercussions of transformative change can be meaningful, reaching well beyond our immediate sphere. For instance, brands like Patagonia and Reformation, through their unwavering dedication to sustainability and ethical practices, have cultivated devoted followings and prompted a reevaluation within the larger realm of the fast fashion apparel sector. Likewise, the #MeToo movement germinated from individual narratives and burgeoned into a worldwide wave, precipitating tangible shifts in societal norms and across corporate policies.

Both individual and corporate transformations, particularly in AI, are deeply intertwined, thriving on growth, discovery, and the pursuit of excellence. The impact of transformative change, much like the advancements in AI, can be far-reaching, extending beyond our immediate sphere. Consider, for instance, how AI-driven companies are redefining industries and influencing societal norms, similar to how brands like Patagonia have transformed the fashion industry with their commitment to sustainability. Just as the #MeToo movement grew from individual stories and brought about global change, AI—rooted in ethical practices and human-centric values—is catalyzing a reevaluation of business processes and societal interactions.

As Logan Gelbrich succinctly puts it: "The best version of you is on the other side of the proverbial mountain. The journey will require effort and adversity." (Gelbrich n.d.).

Activity:

Can you recall a change in your life that had ripple effects on those around you or vice versa? How did this change impact your personal or professional journey?

Conclusion

The journey through data and AI is both a continuation and an expansion of the transformative path. Embracing a growth mindset empowers the effective leveraging of data and AI, encouraging a culture of continuous learning, agility, and adaptability. This mindset prepares us to face the challenges and uncertainties of the digital age, turning obstacles into opportunities for growth and innovation.

The principles and insights shared in this chapter serve as a foundation for navigating the future of personal and organizational growth. The journey does not end here; it evolves, inviting the further exploration of applications, strategies, and the potential of data and AI to drive transformative change. A growth mindset and a deep understanding of data and AI are poised to create a future that is not only digitally advanced but also profoundly human-centric and sustainable.

CHAPTER 3

Harnessing the Power of Data through Insights and Decision-Making

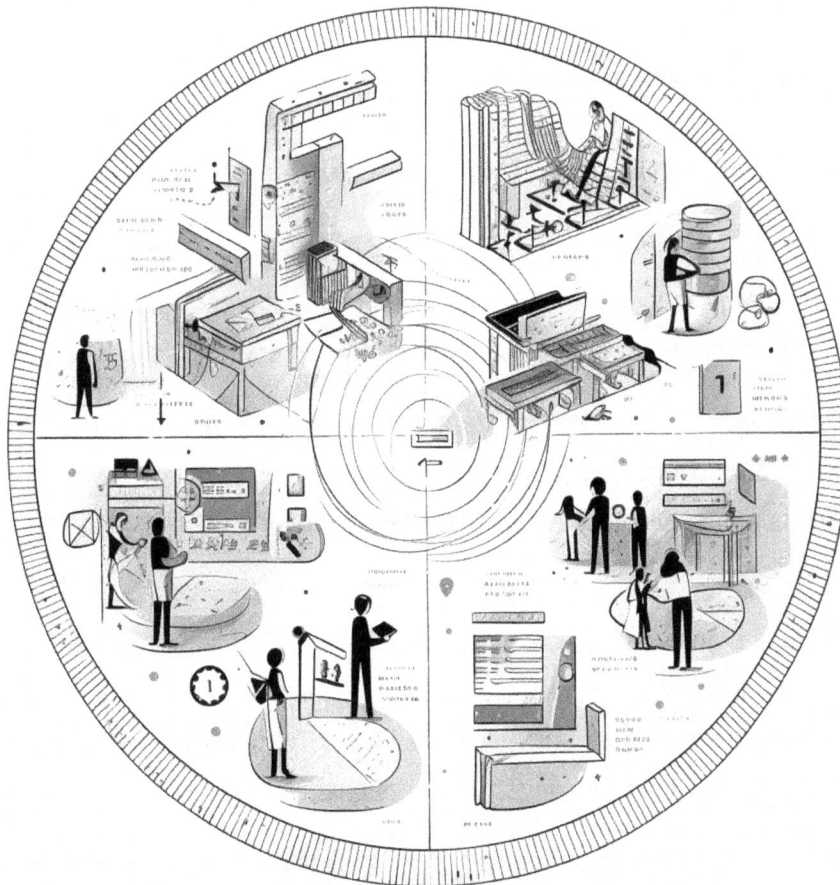

What if I told you that the most valuable asset when transforming a business is data, not a product or service?

Ha grappled with the challenges and reaped the rewards of a data-driven approach. Once, at my startup, we faced a significant data challenge, with our sales data scattered across various spreadsheets that made it nearly impossible to make accurate forecasts. We had a powerful vision—but lacked the insights to make informed choices to propel us forward. We could not harness our data's potential without a centralized data system. Venturing forward, we began a transformative journey.

In those early days, we faced significant resistance within our company culture toward data-driven decision-making. We relied on incomplete and messy data to make product prioritization decisions. However, when we invested in data pipelines and validation, we uncovered how flawed our initial assumptions were, learning the critical importance of data quality and integrity. Despite our familiarity with KPIs, we struggled to trust data sources due to our weak data foundation.

This led us to confront the data silos that were preventing us from accessing the information we needed to drive decision-making. Data silos, essentially, are like isolated islands of information stored separately within an organization. They prevent the seamless flow and access of data across different departments. This fragmentation was not just a hurdle; it was like carrying extra weight when racing more agile competitors.

Addressing Challenges in a Data-Driven Culture

Implementing a data-driven culture comes with several challenges, which can be addressed as follows:

- **Data silos**: Information is often fragmented across different departments, making it difficult to gain a holistic view of the organization. To address this, implement a centralized data platform that integrates material from various sources, enabling seamless access and analysis.

- **Data quality and accessibility**: Inconsistent or poor-quality data can lead to inaccurate insights. Establish robust data governance practices, including data cleaning, validation, and standardization. Ensure that data is accessible to those who need it by creating user-friendly dashboards and reporting tools.

- **Cultural resistance**: Employees may resist changes associated with a data-driven approach, especially if it challenges established practices. Overcome this by fostering a culture of transparency and collaboration. Provide training to enhance data literacy and involve employees in developing data-driven strategies, making them feel invested in the change.

- **Ethical concerns**: The use of data, especially in AI applications, raises ethical issues, such as bias and privacy. Implement ethical guidelines and governance frameworks to ensure that data is used responsibly. Regularly audit AI systems to detect and mitigate biases and ensure data usage complies with legal and ethical standards.

In several of the AI projects I have led, particularly in marketing and sales, I have encountered the crucial need to balance the raw power of technology with the nuanced insights that only human expertise can provide. This realization became particularly evident when we were automating cumbersome processes that traditionally relied on the deep knowledge of domain experts with years of experience.

One example that stands out is when we decided to automate aspects of our marketing strategy. While AI could quickly analyze vast amounts of data and suggest optimal strategies, it became clear that the most effective solutions emerged when we incorporated the expertise of our marketing team. These experts had a profound understanding of our audience—something that AI, with all its data-crunching power, could not fully replicate. To bridge this gap, we began modeling their input into a knowledge graph, which allowed us to structure and integrate these insights into our AI systems. However, we soon realized that this was not enough. The real breakthrough came when we recognized the importance of keeping humans in the loop as overseers and active participants in the AI process.

As we started testing our solutions, we noticed that users—both our team members and our clients—provided invaluable feedback. This feedback often highlighted nuances that the AI had missed or misinterpreted. For instance, a recommendation might be technically correct but did not account for the subtle cultural preferences of a specific market segment. We realized that human input was crucial for refining our AI models and ensuring our solutions were practical and applicable in real-world scenarios.

To effectively capture feedback, we implemented reinforcement learning approaches, allowing our AI systems to learn and improve from human corrections and insights. By incorporating this continuous feedback loop, we updated our knowledge graph and enhanced our AI solutions, ensuring they remained adaptive and relevant.

This experience solidified my belief that the goal of AI should not be to replace humans but to empower them. By automating routine processes and decision-making tasks, we aimed to free up human creativity—allowing our teams to focus on their work's strategic and innovative aspects without getting bogged down in repetitive tasks.

This kind of data fragmentation is a challenge I have encountered firsthand repeatedly. At an e-commerce startup I worked with, customer data was fragmented across various systems tied to individual business units. When leadership asked for a consolidated customer view, providing accurate data sources took much work to truly leverage the customer data, and breaking down those silos became a top priority. It took immense effort and time to connect these scattered data sources and then structure the data in ways that extracted meaningful insights.

Building on this experience, we began to unlock the data's actual value for the organization. Raw data was gradually connected like scattered puzzle pieces as we meticulously structured and organized it. This evolution underscored the transition from data to information, then to knowledge, and on to wisdom, a process that required infusing meaning, context, and interpretation into the data.

We turned to the Google Cloud Platform (GCP) ecosystem, a pivotal tool to transform scattered data into actionable wisdom. By 2018, our choice of tooling within GCP was strategic, focusing on integration and scalability. We leveraged Google BigQuery for its powerful data warehousing capabilities, allowing us to quickly and efficiently analyze vast datasets. Google Cloud Storage provided a secure and scalable solution for storing raw data, while Google Dataflow seamlessly processed and transformed our information in real-time. This setup streamlined the data-to-insight pipeline and infused our raw data with context and meaning, guiding the transition from mere data to valuable knowledge and wisdom.

Reflecting on the startup's five-year evolution, a transformative shift was recognized: data transitioned from a byproduct to a cornerstone of strategizing. This realization came when the buzz around the data's potential gained momentum. Direct experience underscored its impact—data-driven insights streamlined operations, curtailed expenses, and bolstered profitability. Leveraging tools like BigQuery and Dataflow was a game-changer. These tools enabled us to analyze and manage data effectively and highlighted the strategic importance of data in real-time decision-making and long-term planning.

I have worked with various companies for over two decades. Interestingly, from startups striving to gain a competitive edge to established businesses aiming to stay ahead, the potential of data-driven transformation is undeniable. The data-driven approach enables the optimization of processes, which results in higher efficiency and profitability. An example is optimizing delivery routes to reduce fuel consumption and improve delivery times by analyzing routes and transportation data. Improvements led to cost savings and enhanced customer satisfaction, positioning a logistics company as a reliable partner.

Unlocking the value of data presents manifold challenges. Despite recognizing that doing so is a strategic asset, many companies need help extracting meaningful insights. However, organizations can overcome these challenges and effectively harness data's power.

Case Studies

Netflix: Evolution and Adaptation

- **Challenge**: From its origins as a DVD rental service to becoming a streaming behemoth, Netflix's challenge has been maintaining viewer retention and engagement amidst fierce competition (Brand Credential 2023).

- **Data-driven insight**: By the mid-2010s, Netflix was pioneering data-driven content recommendation algorithms. They utilized viewer data to personalize experiences, fundamentally transforming the streaming landscape. This strategic shift enhanced user engagement and was pivotal in reducing churn.

- **Competitive landscape**: As Netflix transitioned to streaming in the late 2000s, competitors like Hulu and Amazon Prime Video emerged, each vying for market share. The streaming wars intensified in the 2010s with the entry of Disney+ and Apple TV+, making the need for data-driven decision-making more critical (Chong 2020).

- **Outcome**: Using data analytics for personalized content recommendations significantly decreased their churn rate and cemented their market dominance (Natalie 2017).

Uber: Data-Driven Efficiency

- **Challenge**: Uber, established in 2009, faced the challenge of optimizing its ride-matching process to reduce wait times as it expanded globally.

- **Data-driven insight**: By the mid-2010s, Uber refined its algorithms using real-time data to ensure quicker pickups and an improved rider experience. This was key to staying ahead as the ride-hailing market became crowded with competitors like Lyft.

- **Competitive landscape**: In an industry rapidly embracing technology, Uber's data-centric approach to optimizing routes has become a critical differentiation, ensuring customer loyalty and operational efficiency in an increasingly competitive space (Dobler 2022).

- **Outcome**: Uber's refinement of its matching algorithms through data analysis reduced wait times and improved rider satisfaction (Sawhney et al., 2020).

Amazon: Pioneering Personalization in E-commerce

- **Challenge**: As Amazon grew from an online bookstore to a vast e-commerce platform, the challenge was to personalize the shopping experience to increase sales and customer retention.

- **Data-driven insight**: By the late 2000s, Amazon was leveraging its vast trove of customer data to create a sophisticated recommendation engine, providing personalized product suggestions that increased engagement and loyalty.

- **Competitive landscape**: With the rise of e-commerce platforms like eBay and, later, Walmart's online foray, Amazon's commitment to a data-driven, customer-centric approach allowed it to remain a leader in the space, even as the landscape grew more competitive (Mpeshev 2023).

- **Outcome**: Amazon's ability to increase sales and customer loyalty through personalized recommendations showcases its data prowess.

Examining these companies reveals the transformative power of data-driven insights. Even in the face of challenges, harnessing data effectively can lead to breakthroughs and significant competitive advantages. These stories underscore the importance of a robust data foundation, the ability to extract meaningful insights, and the agility to act on those insights. Whether you are a budding startup or an established enterprise, the lessons from these companies highlight the undeniable potential of data-driven transformation.

Data and AI as a Universal Solution Regardless of any Size

Every organization faces unique challenges, whether a nimble startup or an established brand. Startups may navigate limited resources and tight budgets, where each decision could pivot the company's future.

Conversely, established enterprises manage vast resources to innovate and stay ahead in a competitive landscape. Both must leverage data, AI, and computing to refine strategies and maintain a competitive edge.

Interestingly, the challenges startups face are not entirely dissimilar to those of more extensive, well-established businesses. Despite varied sizes and histories, certain obstacles remain constant. Whether a small startup or a global corporation, the strategic use of data and emerging technologies can make a significant difference in achieving success.

One advantage of startups over larger organizations is their ability to adapt and transform quickly. With a more agile structure and a willingness to embrace change, startups can pivot strategies quickly while embracing data-driven decision-making. This flexibility allows them to capitalize on emerging opportunities and stay ahead of the curve.

Whether you are a budding entrepreneur launching a startup or a seasoned professional in a well-established organization, leveraging the power of data, AI, and computing is a force multiplier.

The DIKW Pyramid: Unveiling the Value of Data

A foundational framework in digital transformation, the data, information, knowledge, and wisdom (DIKW) pyramid is critical in data contexts and plays a crucial role in other fields, such as information management, knowledge management, and information systems. Since its inception, it has contributed to our understanding of data transformation, with crucial contributions from scholars like Milan Zeleny and Russell Ackoff.

Zeleny's work, particularly his book *Human Systems Management: Integrating Knowledge, Management, and Systems*, explores the integration of data and knowledge within organizational contexts, providing a comprehensive view of how information evolves into actionable wisdom. Ackoff's seminal paper "From Data to Wisdom" appeared in the *Journal of Applied Systems Analysis* and lays out the conceptual framework of the DIKW hierarchy, illustrating the progression from raw data to insightful wisdom.

The pyramid begins with data, which are discrete facts lacking context. As this data is structured and contextualized, information (such as customer profiles) is derived that provides a deeper understanding of preferences and behaviors. Beyond that, knowledge is developed to inform actions and strategies when information is further analyzed and synthesized. At the apex is wisdom, which requires judgment and the application of knowledge to make ethical decisions and to pursue ideals—a uniquely human attribute.

This model underlines that every level of the pyramid builds upon the previous one, signifying the progressive enrichment of raw data into actionable wisdom (Frické 2018).

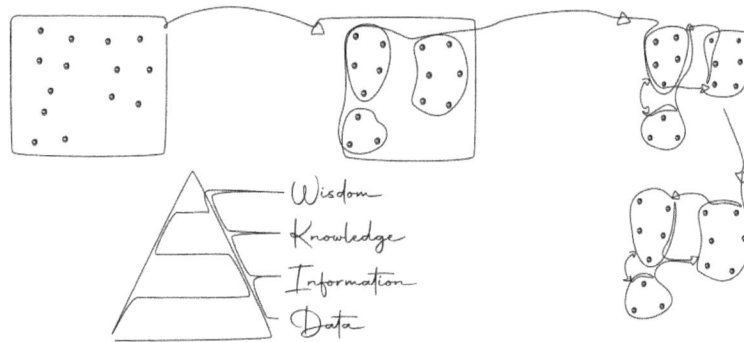

Figure 11. From Data to Information, Knowledge, and Wisdom

Transitioning from data to information, knowledge, and wisdom involves adding meaning, context, and interpretation. Knowledge is attained when information is imbued with understanding, enabling informed decision-making. Connecting data sources with knowledge graphs enhances understanding of things like customer profiles. Additional knowledge, such as customer segments based on interactions, sales data, and marketing insights, can be inferred. Such knowledge empowers organizations to make informed decisions and take appropriate actions.

Wisdom is achieved when organizations leverage understanding to create new knowledge and make proactive decisions. By leveraging customer profiles and segment knowledge, organizations can approach buyers and sellers with personalized recommendations, fostering innovation, optimizing operations, and enhancing customer experiences.

Data as a Strategic Asset

Data has transcended its status as a mere byproduct of business operations and has become a strategic asset. Organizations that can effectively leverage it gain a significant competitive advantage. Let us explore some examples of how data acts as a strategic asset.

Figure 12. Data Creates Value

Higher Efficiency and Profitability

Data-driven insights can optimize processes and reduce costs, increasing efficiency and profitability.

- **Example**: UPS has seen substantial benefits from implementing its ORION system, which optimizes delivery routes. Since fully deploying it in the United States, UPS has reduced its annual distance driven by about 100 million miles and cut fuel consumption by 10 million gallons. This means avoiding 100,000 metric tons of greenhouse gas emissions each year, alongside cost savings of $300-400 million (UPS 2020).

Furthermore, each daily per-driver mile reduced is estimated to save UPS $50 million annually, highlighting the significant impact that even small efficiencies can have when scaled across the entire fleet (Chegg, n.d.).

Optimized Customer Experience

Data enables organizations to understand customer preferences, behavior, and sentiment, leading to personalized and optimized customer experiences.

- **Example**: Netflix's recommendation system contributes significantly to its success, driving over 80 percent of the content streamed by subscribers. This level of personalization has enabled Netflix to excel in customer retention, boasting a rate of over 90 percent compared to Hulu's 64 percent and Amazon Prime's 75 percent.

The recommendation engine uses a variety of data points from subscriber interactions to create detailed profiles and offer customized content, thereby reducing the need for extensive marketing and advertising. These strategies have been pivotal in helping Netflix achieve and maintain its position as a leader in the streaming industry. For a comprehensive study on Netflix's use of analytics, refer to the article on Markivis' website (Netflix n.d.; Ravulakollu 2023).

Rapid Speed to Market

By harnessing data, organizations can gather market insights, identify emerging trends, and respond quickly with new products or services. Fast-fashion companies analyze real-time sales data, social media trends, and customer feedback to rapidly design and launch new fashion collections, staying ahead of competitors and meeting customer demands.

- **Example**: Zara has built an operation that responds rapidly to real-time data analytics. The company's unique business model involves keeping most production in-house and local. It can move a garment from design to store in just a few weeks, delivering new items twice a week to their stores—much faster than the industry standard. Zara's manufacturing proximity allows for rapid product turnover, meaning it can quickly capitalize on emerging trends. This strategy results in limited production runs for new items, creating an environment of scarcity and demand (Magsino, n.d.).

Regulatory Compliance

Data helps organizations ensure regulatory compliance by providing accurate and auditable records. Financial institutions use data analytics and reporting tools to monitor transactions, detect anomalies, and comply with anti-money laundering regulations, mitigating compliance risks and avoiding penalties.

- **Example**: JPMorgan Chase's distinctive use of AI is its technology and large-scale applications across global operations. It is invested substantially in AI and ML, as demonstrated by a 2023 commitment of over $1.5 billion in AI and ML across numerous use cases. The firm's AI/ML strategy and significant technology budget allow for analyzing transactions at an unprecedented scale, employing AI not just for detecting fraudulent activities but also to provide personalized banking services, improved customer service, and optimized internal operations, all beyond the typical regulatory compliance operations (JPMorgan Chase n.d.).

Enablement of New Products and Services

Data fuels the development of innovative products and services, opening doors to new business models.

- **Example**: Spotify's Discover Weekly is a standout example of utilizing user data to craft personalized experiences that resonate deeply with individual preferences. Introduced in 2015, the playlist compiles songs based on each listener's unique musical tastes, harnessing ML to analyze streaming history, user-created playlist data, and even the specific audio attributes of songs listened to. By dissecting elements like beats, keys, and dynamics, Spotify's algorithms offer a tailored set of weekly tracks, driving significant user engagement. This feature has seen users stream billions of hours, discovering new songs and contributing to the exposure of emerging artists globally.

Moreover, the Discover Weekly tool showcases Spotify's commitment to data-driven personalization, which has had broader implications in influencing music production and artist promotion strategies. The playlist curation goes beyond mere suggestions; it fosters an environment where users are consistently introduced to music that might have remained undiscovered, creating a dynamic and ever-evolving listening experience (Pasick 2015).

Dilemma: Unlocking the Full Potential of Data

Despite the immense possibilities of data, many organizations need help to realize its full potential. A survey conducted by Gartner (Gartner 2023) reveals that less than half of data and analytics leaders believe their teams effectively provide value to their organizations. Moreover, only a tiny percentage perceive data and analytics projects as generating tangible value or demonstrating superior financial performance. These findings underscore the gap between the promise of data and its implementation.

The McKinsey report "The Data-Driven Enterprise of 2025" emphasizes the need for organizations to prioritize diligence in data-driven initiatives, develop comprehensive data strategies, ensure data quality and accessibility, and foster a culture that values data-driven decision-making. The report highlights how rapidly advancing technology and increasing data literacy transform what it means to be data-driven while underscoring the critical role of integrating data into every decision, interaction, and process within an organization (QuantumBlack: AI by McKinsey, 2022).

MIT Sloan highlights the challenges of gaining an analytics edge and fully leveraging data, emphasizing the importance of investing in data infrastructure, developing data literacy across the organization, and embracing advanced analytics techniques (Stackpole, 2023).

> *"The goal is to turn data into information and insight."*
> **— Carly Fiorina, former CEO of Hewlett-Packard.**

The dilemma of organizations in unlocking the full potential of data requires a strategic approach that includes robust data governance, investment in data literacy and talent, adoption of advanced technologies, and cultivating a data-driven culture.

By effectively addressing the challenges outlined previously, organizations can bridge the gap between data's potential and its actual benefits, which derive from harnessing insights to drive innovation, streamline operations, and secure a competitive edge.

Challenges in Unlocking the Value of Data

Despite recognizing data as a strategic asset, many companies need help to extract meaningful insights and value from it. I experienced this firsthand when my company rapidly expanded its business, outpacing our data capabilities. We faced issues such as inventory mismanagement and inefficient operations, highlighting the need for a centralized data strategy.

For example, our inventory tracking systems failed to keep pace during rapid growth, leading to frequent stockouts and overstock situations. We implemented a robust data management system that integrated inventory data across all departments to address this. This system provided real-time insights into inventory levels, helping us optimize stock and improve operational efficiency.

This experience reinforced that data is not just about numbers but about aligning those numbers with a vision and strategy to drive meaningful impact. By defining a clear data strategy and aligning it with our business goals, we transformed our operations and achieved significant improvements in efficiency and accuracy.

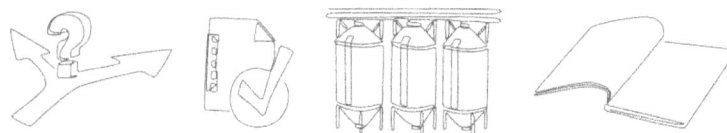

Figure 13. Businesses and Organizations Failing to Get Value from Data

Breaking down these data silos was not just a technical challenge; it was a lesson in perseverance. I learned that while tools and technologies are essential, the real breakthroughs come from relentless curiosity and the willingness to challenge the status quo. This experience reinforced my belief that effective data management is as much about mindset as technology.

Data Quality and Integrity

Data quality can lead to reliable insights and sound decision-making. Robust data governance practices, including validation and ongoing monitoring, are crucial for ensuring quality and integrity.

- **Example**: A notable incident involved JPMorgan Chase, where they faced an FCA fine for failing to report traded derivative transactions between 2007 and 2015. This breakdown in data governance led to significant regulatory repercussions, highlighting the real-world impact of inadequate data management (Day 2013).

Data Silos and Fragmentation

Data fragmentation across systems, departments, and platforms hampers the ability to connect and integrate data from different sources, limiting the holistic view necessary for comprehensive insights. Breaking down data silos and establishing a unified data ecosystem is critical for unlocking the full potential of data.

- **Example**: Amgen faced challenges storing data across multiple legacy systems, which created barriers to efficient drug development due to difficulties accessing and analyzing information. Amgen integrated disparate data sources and streamlined its research process by transitioning to a unified data platform and leveraging lakehouse architecture through Databricks.

The shift allowed Amgen to improve data ingestion rates, cut processing times by 75 percent, and deliver insights twice as fast—all while reducing compute costs (Dominic & Johnson, 2022).

Lack of Data Literacy and Skills

Data literacy among employees is essential for organizations to derive insights and effectively make data-driven decisions. Investing in data literacy programs and upskilling initiatives can address this barrier and equip employees with the necessary skills.

- **Example**: Merck, a global healthcare company, implemented an upskilling and data literacy program early in their data strategy to ensure all employees—and not just those on the data team—recognized the value of data. This initiative was about teaching technical skills and fostering a culture where data was seen as an asset across the organization. By emphasizing data literacy and providing relevant training, employees could see how data delivers value in their roles, which led to more data-centric company culture (Nehme, 2023).

Insufficient Data Infrastructure

Inadequate data infrastructure can impede an organization's ability to handle and process large volumes of data. Investing in scalable and modern data infrastructure, including cloud-based solutions, is crucial for effectively handling data and deriving valuable insights.

- **Example**: By migrating to Amazon Web Services (AWS), Netflix significantly enhanced its infrastructure's scalability and reliability. With AWS, Netflix could quickly deploy servers and storage to support a seamless global service. This strategic move to scalable cloud infrastructure enabled Netflix to handle large volumes of data and streaming requests without interruption, effectively supporting surges in user activity and providing a consistent viewing experience (AWS 2016).

Data Privacy and Security Concerns

Striking the right balance between data utilization and confidentiality is essential for maintaining customer trust. Implementing robust data security measures and adhering to privacy best practices is crucial to unlocking data values while protecting customer privacy.

- **Example**: The 2018 Facebook and Cambridge Analytica scandal involved misusing millions of Facebook users' data, harvested without consent. It led to global outrage and calls for tighter regulations. The case exposed the severe implications of unethical data practices. It underscored the need for rigorous adherence to privacy principles to maintain user trust and comply with legal standards (Facebook–Cambridge Analytica data scandal 2024).

Lack of Clear Data Strategy

Organizations may struggle to prioritize and align their data initiatives with business goals without a clear data strategy. Developing a comprehensive data strategy enables organizations to focus their efforts and investments on the most valuable areas.

- **Example**: The retail giant Walmart improved its operations by employing collaborative planning, forecasting, and replenishment (CPFR) and vendor-managed inventory (VMI), which allow for a highly integrated and efficient supply chain. These strategic implementations, facilitated by advanced information technology, enabled Walmart to collaborate closely with suppliers, streamline inventory management, and ensure product availability. The outcome improved efficiency in inventory management, cost savings, and a better customer experience due to reduced stockouts and more competitive pricing (MBA Knowledge Base n.d.).

Cultural Resistance to Data-Driven Decision-Making

Organizational culture plays a significant role in the successful implementation of data-driven decision-making. Overcoming resistance to change and fostering a culture that values evidence-based decision-making is crucial for driving progress.

- **Example**: While steeped in engineering excellence, traditional automakers initially hesitated to adopt data-driven design principles that Tesla had embraced at its inception. Tesla's approach involved leveraging Silicon Valley's culture of "techpreneurship," a clear break from traditional automotive design methods, enabling it to innovate rapidly and disrupt the industry. By adopting a tech company model for vehicle production and system development, Tesla could implement advanced features, like over-the-air software updates for vehicle recalls, giving them a significant market advantage. This showcases the power of data-driven strategies to foster innovation and stay ahead of market trends (Rassweiler & Brinley 2014).

By recognizing and addressing these challenges, organizations can overcome barriers and fully harness the power of data.

Activity:

Think about your organization's current data practices. Which of the challenges discussed in this chapter resonate most? What steps can you take to address them?

Enhance Data Strategies: The Free Data Maturity Assessment Tool

I am thrilled to present the Data Maturity Assessment Tool, accessible for free on my website. This comprehensive, user-friendly tool is designed to measure an organization's data-driven readiness and pinpoint precisely where it stands on the path to data mastery. This tool benchmarks current data practices against industry standards and unlocks bespoke, actionable insights to propel data initiatives forward. **By leveraging this tool, you will:**

- Receive a personalized data maturity profile.
- Identify specific areas ripe for improvement.
- Obtain custom recommendations tailored to your business's unique context.
- Start a journey towards data-driven excellence with clear, achievable steps.

It is a zero-cost investment in your organization's future. Visit the Data Maturity Assessment Tool now to take the first step towards transforming your data practices and charting a course toward unparalleled data proficiency.

Conclusion

While the potential of data and AI is vast, realizing its full transformative power requires overcoming significant obstacles—from ensuring data quality and breaking down data silos to fostering data literacy and developing a comprehensive data strategy.

Adopting a data-driven mindset and continually evolving ways to leverage data and AI effectively is essential. My focus now shifts from the foundational understanding and application of data and AI to a deeper dive into the heart of transformation through insights and decision-making.

CHAPTER 4

Exploring the Impact of AI

In personal transformational journeys, our senses become heightened and more attuned to the nuanced signals life is sending us. This enhanced awareness allows us to absorb and interpret the world more deeply, transforming insights into wisdom. Analogously, AI represents a technological leap and a profound expansion of our collective perceptual toolkit. It mirrors the quest for deeper understanding by sifting through data layers with exceptional insight.

AI embodies the principle of extended perception, drawing patterns and insights from areas of digital information that can often seem as vast and enigmatic as the cosmos. It does not just analyze data—it translates it into a language that can be acted upon, offering a bridge between the seen and the unseen, the known and the yet-to-be-discovered.

AI catalyzes the decision-making process, echoing cognitive abilities to synthesize and evolve. AI is more than algorithms and computation; it is a partner in a journey toward enlightened leadership and conscious business practices. AI's evolution can parallel personal growth paths, illuminating the way forward with insights forged from the intersection of data and intuition.

A Glimpse into AI's Past

The development of AI is a fascinating tale spanning several decades. Its roots can be traced back to the mid-20th century when the term "artificial intelligence" was first coined. Early pioneers like Alan Turing, who proposed the idea of a "universal machine" that could simulate any human intellect, laid the groundwork for what would become a revolutionary field (Peralta 2002).

Researchers first began exploring the concept of machines that could exhibit human-like intelligence in the 1950s. In the 1960s and 1970s, AI research was primarily rule-based, focused on creating systems that could mimic certain aspects of human intelligence by following explicit programming rules. These early systems did have practical applications despite their limitations. For instance, the domain of expert systems emerged during this era, and AI was used to emulate the decision-making abilities of human experts in fields such as medicine and geology.

These systems, rudimentary by today's standards, managed to perform tasks such as diagnosing illnesses or recommending mineral exploration strategies based on programmed expertise. However, they were limited in their adaptability and scalability, mainly due to the finite nature of rule-based logic and the computational power available at the time.

The transition from theoretical to real-world applications gained significant momentum with the advent of machine learning in the 1980s, which shifted the focus from manually programmed instructions to algorithms that could learn from data. This was the era when expert systems were becoming increasingly popular. These systems used a "knowledge base" of facts and a set of rules to infer new facts or make decisions, which marked the beginning of AI's journey from the lab to the living room. It began to influence everything from industrial processes to personal devices, setting the stage for the intelligent systems we interact with today.

The 1990s and early 2000s marked a shift towards data-driven approaches. With the advent of the Internet and the digital age, a data explosion occurred. This abundance of data and advances in computational power propelled ML to the forefront, enabling AI to learn from data and improve its performance. This early progress paved the way for a significant milestone in AI advancement—the emergence of deep learning.

Figure 14. AI Did Not Become Disruptive Until Recently.

Deep learning is a subset of ML that leverages artificial neural networks inspired by the structure and function of the human brain. Imagine a vast web of interconnected nodes, much like neurons in the human brain, which take in data, process it through multiple layers of nodes, and produce output, such as recognizing an image or understanding spoken words.

These deep learning algorithms, powered by significant improvements in computing capabilities, revolutionized AI by enabling it to process and analyze complex data with exceptional accuracy. This breakthrough opened new possibilities in areas such as image recognition, natural language processing (NLP), and speech recognition.

- **Advantages**

    o **Adaptability**: Neural networks learn from data, which means they can adapt to changes, making them invaluable for constantly evolving industries.

    o **Efficiency**: Tasks that once took humans hours can now be completed in seconds, from analyzing complex datasets to recognizing patterns in customer behavior.

    o **Innovation**: With neural networks, businesses can offer new, cutting-edge services and products that set them apart from competitors.

- **Challenges**

    o **Data needs**: Neural networks require vast amounts of data to learn effectively, and having the correct data is crucial.

    o **Implementation costs**: Setting up neural networks can be resource-intensive initially, though the long-term benefits often outweigh the costs.

    o **Ethical considerations**: As with all AI, there are moral implications, especially concerning data privacy and potential biases in decision-making.

Building upon the breakthroughs of deep learning, the latest frontier of AI innovation is generative AI.

While AI has been making significant strides, recent advancements in GenAI have genuinely brought about disruptive change. Generative AI refers to AI models and algorithms that can create new content or perform new tasks with little or no explicit training. This marks a shift from traditional AI systems that rely heavily on pre-programmed rules or extensive training datasets. While we have seen significant advancements and applications in the present, the future holds even more promise for AI. We can envision new capabilities and opportunities as AI propels rapid advancements across sectors.

Real-World Applications

Artificial intelligence is revolutionizing various sectors in unprecedented ways, but underlying many of these advances is a lesser-recognized hero: AI in analytics. It is an invisible engine of digital transformation that powers real-world applications across multiple industries.

Data is not just an asset but the lifeblood of organizational functionality. AI quickly sifts through this voluminous data, providing insights, predictions, and optimizations that transform traditional workflows. Here are some of AI's analytical impacts across various sectors.

Healthcare

- **Successes**

    - **Disease prediction and prevention**: AI algorithms analyze patient data to predict potential diseases before they manifest, allowing for early intervention. Predictive analytics are essential here.

    - **Real-world business outcome**: Healthcare institutions use AI-powered analytics to forecast patient risks and allocate resources more effectively.

    - **Medical imaging**: Deep learning models assist radiologists in detecting anomalies in X-rays, MRIs, and CT scans with higher accuracy.

    - **Drug discovery**: AI accelerates drug discovery by analyzing complex biochemical interactions.

- **Challenges**

    - **Sequestration**: Data privacy concerns and the need for anonymized patient data.

    - **Integration**: Incorporating AI tools into healthcare workflows can take time and effort.

Finance

- **Successes**

    - **Fraud detection**: AI systems can detect unusual patterns and activities, helping to detect fraudulent transactions early. In this case, real-time analytics are invaluable.

- **Real-world business outcomes**: Fintech companies save millions annually through immediate fraud detection using AI analytics.

- **Robo-advisors**: Automated platforms provide online financial advice or portfolio management with minimal human intervention.

- **Challenges**

  - **Oversight**: The financial sector is heavily regulated, and there is a need for transparent AI models.

  - **Too much free will**: Concerns about AI making autonomous decisions that can affect market dynamics.

Entertainment

- **Successes**

  - **Content recommendation**: Platforms like Netflix and Spotify use AI to analyze user preferences and recommend content. Predictive analytics also helps identify potentially successful content pieces.

  - **Real-world business outcome**: Increased customer engagement and maximized revenue through personalized recommendations.

  - **Video games**: AI dynamically enhances the gaming experience by adjusting real-time difficulty settings to match a player's skill level. This intelligent adaptability ensures that novice and veteran gamers are equally challenged and entertained, leading to a more engaging and personalized experience.

- **Challenges**

  - **Too much focus**: More reliance on algorithms can sometimes lead to narrow content exposure for users.

  - **Harmful content**: Ethical concerns about deepfakes and the potential misuse of AI-generated content.

Agriculture

- **Successes**

  - **Precision farming**: AI-driven drones monitor crops and analyze data to optimize irrigation, pest control, and harvesting.

  - **Crop prediction**: ML models predict crop yields, helping farmers plan their harvests better.

- **Soil health monitoring**: Advanced AI algorithms interpret data from soil sensors to assess health indicators such as moisture, nutrients, and pH levels. This real-time monitoring allows for tailored soil management practices, ensuring optimal growth conditions and sustainable farming practices.

- **Challenges**

  - **Capital intensive**: AI-driven agricultural tools require high initial investment.

  - **Learning curve**: Training farmers to integrate and use AI tools effectively is required.

Retail

- **Successes**

  - **Supply chain optimization**: AI predicts demand, helping retailers manage inventory efficiently.

  - **Personalized shopping experience**: Online retailers use AI to provide personalized shopping recommendations based on user behavior.

- **Challenges**

  - **Quarantining personal data**: Concerns about data privacy and how consumer data is used for personalization.

  - **Unifying operations**: The challenge of integrating AI into traditional retail models.

In these industries, AI is not just a tool but a transformational force, reshaping business models, workflows, and the products and services offered. However, with its vast potential come challenges that industries must navigate. Balancing the promise of AI with ethical, practical, and economic considerations is crucial in moving towards an even more AI-integrated future.

Envisioning the Next Decade

AI's potential continues to captivate as we stand on the precipice of a new era. While predicting the future is always speculative, current research and expert opinions provide a tantalizing glimpse into the possibilities.

- **Ubiquitous AI**: Just as smartphones became an integral part of our lives in the past decade, AI will become commonplace. From homes to workplaces, AI-driven tools and applications will be at our fingertips, making our lives more efficient and personalized.

- **AI in healthcare**: AI will revolutionize healthcare. From personalized treatments based on an individual's genetic makeup to early diagnosis through pattern recognition, AI will enhance patient care and potentially extend lifespans.

- **Ethical AI**: As AI becomes more integrated into our lives, moral considerations will be more vital. Transparency, fairness, and accountability will be paramount. Regulatory bodies overseeing AI implementations and ensuring adherence to ethical standards might emerge.

- **Collaborative AI**: Rather than replacing humans, AI will work alongside us. In industries like manufacturing, AI-driven robots will collaborate with human workers, each amplifying the other's strengths.

- **AI-driven creativity**: AI will venture into creative domains, assisting in music composition, art creation, and script writing. While the human touch will always be irreplaceable, AI will serve as a tool to enhance and expand creative horizons.

- **Decentralized AI**: With the rise of edge computing and the Internet of Things (IoT), AI processing will move closer to the data source. This means faster real-time decisions and reduced reliance on centralized cloud infrastructure.

- **AI in education**: Personalized learning experiences will become the norm. AI-driven platforms will assess individual student needs and adapt the content in real-time to ensure optimal learning outcomes.

- **AI economy**: Just as the Internet gave rise to entirely new business models and industries, AI will pave the way for innovative economic structures. The increase of AI-driven marketplaces, services, and even new professions is on the horizon.

The next decade promises to be transformative. As AI evolves, it will reshape industries, redefine experiences, and reimagine possibilities. However, with these advancements come responsibilities. Business leaders, policymakers, and society must navigate this journey with foresight, ensuring that the AI-driven future is inclusive, ethical, and beneficial for all.

The Four Pillars of Transformative Power of AI

In the landscape of transformation, understanding the roots and rise of AI is essential, not merely for historical context but to appreciate the full magnitude of its role as an agent of change. The emergence of AI is not just a technological narrative; it mirrors our growth journey—expanding from nascent stages to robust capabilities. This parallel path offers vital insights for conscious-aware leaders seeking to harness AI within their organizations. The transformative power of AI stems from the synergy of several disruptive forces that have sculpted today's AI and symbolize broader themes of innovation and progression.

Figure 15. Four Pillars of AI: Data, Computing, Generative AI, and Knowledge

Four critical disruptions, each a pillar in its own right, **have converged to form the AI era:**

- **Abundance of data**: Our era's digital heartbeat, capturing the human experience and fueling AI's analytical prowess.

- **Computing power**: Vastly improved capabilities allow us to process this data and unlock complex AI algorithms.

- **Machine learning, deep learning, and generative AI**: These techniques enable AI to derive meaningful patterns and predictive insights, similar to the learning and growth process in our lives.

- **Knowledge and know-how**: The democratization of AI education fosters a culture of innovation and shared expertise across borders and industries.

Each of these pillars contributes to AI's current form and resonates with the foundational elements of transformation explored throughout this book. They serve as guideposts for leaders to direct their strategic thinking, integrate technological advances, and lead their businesses toward a future where human and artificial intelligence are merged to pursue progress and purpose.

Pillar 1: Abundance of Data

One of the critical disruptions propelling the AI revolution is the sheer abundance of available data. We generate data in every digital interaction, from online shopping to social media browsing. This data offers invaluable insights into human behavior, preferences, and patterns when aggregated.

Every day, our digital activities contribute to the staggering creation of 2.5 quintillion bytes of data. To put this astronomical number into perspective, imagine this: if a single byte were a grain of sand, then 2.5 quintillion bytes would be more than the entire Sahara Desert. By comparison, a floppy disk—an icon of early personal computing—can hold only 1.44 million bytes (approximately the amount of data needed for a simple text document). In terms of the vast data landscape we are discussing, a floppy disk's capacity is just a minuscule speck—so tiny that it is equivalent to about 0.00000000000576 percent of the data generated daily (Department of Computer Science and Engineering n.d.).

By 2025, it is forecasted that global daily data generation will swell to 463 exabytes. That is nearly a 200-fold increase from 2023, estimated at around 2.5 exabytes daily. This growth is not linear but exponential, reflecting an era where digital interconnectivity is expanding at an unprecedented pace. Such an explosion of data provides a foundational substrate for AI systems to operate. Each byte represents a potential puzzle piece for AI to analyze, learn from, and derive insights that fuel its continuous evolution (Sivarajah et al. 2017).

TikTok exemplifies the vast amounts of data generated by social media. Users spend 52 minutes daily on the platform, and over 1 billion people actively engage monthly. This interaction yields rich information on global human interests and trends (Vardhman, 2024).

By 2021, nearly 60 percent of the global population actively using the Internet has escalated data generation to unprecedented levels, laying the groundwork for AI advancements (Kemp 2021).

Beyond social interactions, the proliferation of IoT devices vividly illustrates data's exponential growth. For instance, smart home systems integrate thousands of sensors that monitor everything from temperature to occupancy, each a node in a vast network feeding AI systems with real-time, actionable data. In agriculture, sensors deployed across farmlands collect soil humidity and nutrient levels, information crucial for precision farming techniques that can significantly enhance crop yields and sustainability.

Data Marketplaces

Managing, sharing, and monetizing this data have evolved simultaneously, evidenced by the rise of data marketplaces. These marketplaces serve as platforms that facilitate the exchange and monetization of data. These platforms are not just data exchanges but innovation hubs where information is the currency driving new business models and strategies.

They connect data providers with valuable datasets with consumers seeking access to specific data for various purposes, such as research, analysis, and AI model training. Data marketplaces operate under the principle that data, like any valuable asset, can be bought, sold, and shared to create mutual benefits.

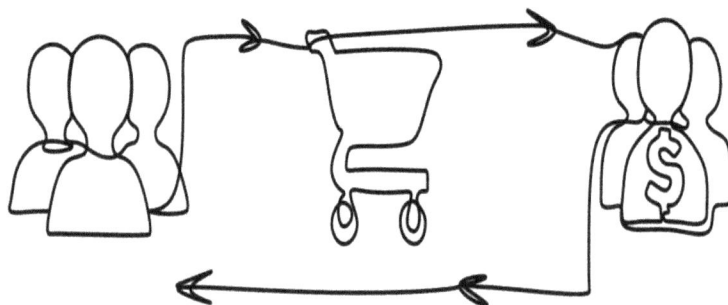

Figure 16. Data Marketplaces Connect Data Providers and Data Consumers

Data marketplaces offer several benefits to both data providers and consumers. Providers can monetize their assets, unlock additional revenue streams, and derive value from untapped data. Consumers gain access to various data sources, accelerating their research, analysis, and AI model development.

However, there are considerations when participating in data marketplaces. Privacy and data governance become critical, as organizations must ensure compliance with regulations and protect sensitive information. Establishing trust and transparency between data providers and consumers is also essential to maintain the integrity and quality of the data exchanged.

As the AI revolution evolves, data marketplaces will play a crucial role in democratizing data access, fostering collaboration, and driving innovation. By embracing internal and external data marketplaces, organizations can tap into the vast potential of data and accelerate their AI-driven transformation.

Internal Data Marketplaces

Within organizations, the concept of an internal data marketplace is gaining traction. Traditionally, data was siloed within departments or teams, making it difficult for various stakeholders to access and utilize information. Internal data marketplaces aim to break down these silos by creating a centralized platform where data is shared across an organization.

Internal data marketplaces enable employees to discover and access datasets from various internal sources. This fosters collaboration, encourages cross-functional insights, and eliminates duplication of efforts. For example, a marketing team might have customer demographic data that the sales team can leverage to refine their targeting strategies. By making this data available through an internal data marketplace, different teams can access and utilize it, enhancing their decision-making processes. Beyond the data, the tools and power we use to process it have evolved.

Pillar 2: Computing Abundance

The evolution of computing power has been nothing short of revolutionary. From the early days of room-sized mainframes to today's pocket-sized smartphones, the trajectory of computational growth has been exponential.

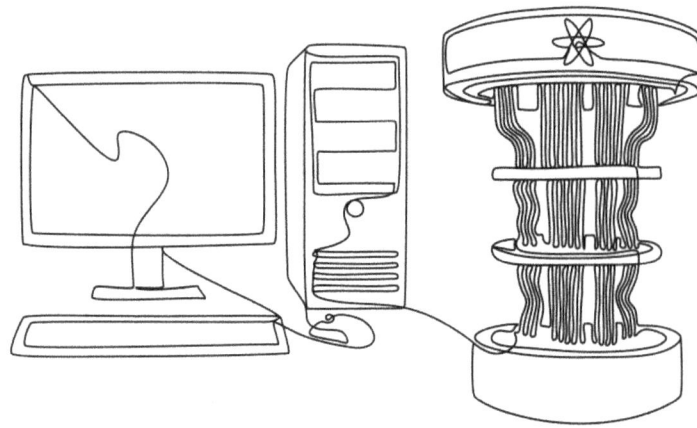

Figure 17. Quantum Computing Brings a New Wave of Disruption

Consider Moore's Law, proposed by Gordon Moore in 1965. This law observes that the number of transistors on a microchip doubles approximately every two years while the cost of computers is halved. This prediction has primarily held, leading to an exponential increase in computing power at decreasing costs (Chojecki, n.d.).

For example, in the 1980s, iconic games like Pac-Man and Tetris had relatively simple 2D graphics and fundamental sound effects due to limited processing power. Although complex for their era, these games were built with a few thousand lines of code. Fast-forward to today, and photorealistic games like Cyberpunk 2077 represent a quantum leap in complexity and immersive experience. While the exact number of lines of code in Cyberpunk 2077 is not publicly disclosed, modern games of similar scope often comprise tens of millions of lines of code and provide advanced graphics, physics simulations, and intricate gameplay mechanics.

This comparison between Pac-Man and Cyberpunk 2077 exemplifies the immense leap in computing capabilities over decades. From the limited but groundbreaking 8-bit era to today's expansive virtual realities, growth is not just in lines of code but also in the sophistication and capability of the software, all powered by the exponential increase in hardware predicted by Moore (Thompson, Ge, & Manso 2022).

As of 2021, the world's most powerful supercomputer was Fugaku in Japan, which could perform over 442 quadrillion calculations per second. This staggering data point is a testament to how computing power has grown exponentially (Burg & Ausubel 2021).

Cloud Computing

Offering fledgling startups and seasoned enterprises a shared platform of vast computational resources, cloud computing has significantly leveled the playing field. Instead of sinking capital into extensive on-premise infrastructure, organizations can now only pay for the computing power they use. This paradigm shift has allowed many organizations to remain agile and responsive to market demands while controlling overhead costs.

For startups, this means the ability to test, develop, and scale applications rapidly without traditional upfront infrastructure costs. Take, for example, a tech startup that develops AI-powered analytics tools. Utilizing cloud services allows them to deploy and test their applications with varying degrees of computational intensity as their user base grows without needing early, substantial investment in hardware.

Established businesses also benefit from shifting their legacy systems to the cloud. A multinational corporation, for example, can use cloud computing to streamline data management across different regions, enhance collaboration, and deploy updates simultaneously worldwide. Cloud platforms also allow leveraging advanced AI and analytics capabilities that might have been prohibitive due to the cost and complexity of setting up a dedicated AI infrastructure.

Quantum Computing

The emergence of quantum computing is redefining the horizon of computing. While still in its nascent stages, this technology has vast disruptive potential. Unlike classical computers that use binary bits (0 or 1), quantum computers leverage quantum bits or qubits, which can exist simultaneously in a superposition of multiple states. This allows quantum computers to process information and tackle specialized tasks quickly.

To underscore this potential, in 2019, Google's quantum computer Sycamore claimed to achieve "quantum supremacy" by performing a task in 200 seconds that would take even the most powerful supercomputers over 10,000 years to execute (Porter 2019).

The abundance of computing power, predicted by Moore's Law, has been a pivotal force fueling AI's rapid progress. With even more processing capacity promised from quantum computing to train complex models on massive datasets, the horizons of artificial intelligence expand exponentially. With the surge in computing power, the methodologies we use to harness this power for AI have also transformed, bringing machine learning to the forefront.

Pillar 3: Machine Learning, Deep Learning, and Generative AI

ML has been the cornerstone of recent AI proliferation. At its core, ML is about teaching machines to learn from data rather than being explicitly programmed by humans. This paradigm shift has enabled machines to tackle previously considered too complex or nuanced challenges.

Consider the domain of image recognition. Traditional programming required developers to write extensive rules to identify objects in images. However, with ML, algorithms are trained on thousands or millions of images, learning to identify objects with remarkable accuracy. Google Photos, for instance, uses ML to automatically categorize and tag photos, recognizing everything from faces to landmarks.

As explained above, the human brain inspires deep learning, a subset of ML. One of its most notable applications is natural language processing. Tools like OpenAI's GPT-3 can generate human-like text, answer questions, write essays, or even create poetry. This is a far cry from the early days of AI when chatbots could only respond to a limited set of predefined queries.

A pivotal case study in deep learning is the game Go. In 2016, AlphaGo, developed by DeepMind, defeated Lee Sedol, one of the world's top Go players. This victory was significant because Go, with its vast number of possible moves, was considered a challenging game for AI. The strategies employed by AlphaGo were not explicitly programmed; instead, it learned from millions of game scenarios, showcasing the power of deep learning.

These advancements in machine learning are not just theoretical; they have practical applications that can reshape industries. AI is already actively transforming a diverse range of industries.

Generative AI Accelerates Disruption

GenAI takes deep learning's capabilities a step further. Instead of just recognizing patterns, generative models can create new, original content. Deepfakes, which use generative adversarial networks (GANs), can produce realistic video footage of real people saying or doing things they never did. On a more positive note, artists and musicians use GenAI to create new forms of art and compositions, blending human creativity with machine precision.

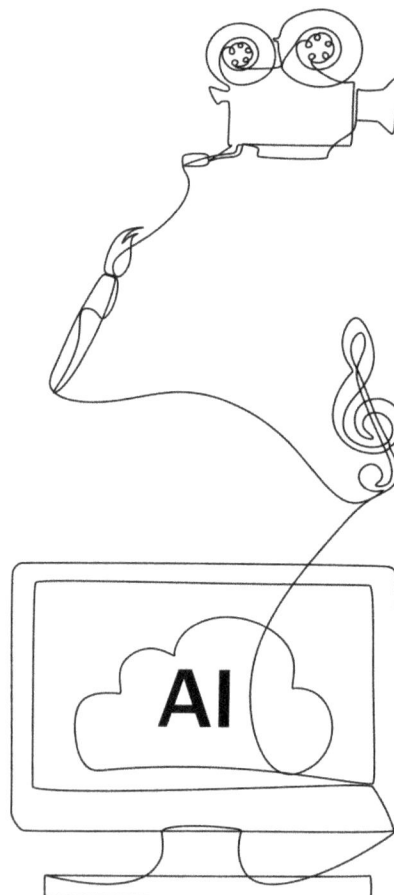

Figure 18. GenAI Is Accelerating AI Disruption

Data Revolution with GenAI

As GenAI's capabilities expand, it is becoming evident that its influence is not limited to direct applications. One of the most profound areas where its ripple effect is being felt regards data, which is creating a revolution that is not just about the sheer volume of accumulated data but also about the innovative ways to create, **manage, and utilize it:**

- **Synthetic data creation**: Tonic.ai is renowned for its advanced synthetic data platform, which is explicitly designed for data scientists. The platform addresses critical challenges such as data scarcity, bias, and access barriers by generating artificial data that mirrors real-world information. This enables comprehensive and accurate training of AI models and facilitates efficient data management, essential for industries ranging from healthcare to financial services (Hemachandran 2023).

- **Illuminating dark data**: GenAI processes and analyzes previously untapped clinical notes and records in healthcare, revealing insights that inform personalized patient care strategies. This capability enables healthcare providers to utilize the full spectrum of available data for better outcomes.

- **Critical data elements (CDEs) identification**: Financial services firms apply GenAI to sift through vast datasets, identifying and prioritizing CDEs for risk assessment. This process facilitates more targeted compliance and decision-making strategies.

- **Data quality and pipeline automation**: DataRobot employs GenAI to automate feature engineering and data preparation tasks, significantly enhancing the accuracy of predictive models. This process improves the reliability of data-driven decisions and accelerates the deployment of AI solutions across various industries (DataRobot n.d.).

Computational Revolution with GenAI

The emergence of quantum computing holds immense promise for solving complex computational problems more efficiently and rapidly. GenAI complements the rise of quantum computing, aiding in analyzing intricate datasets and simplifying quantum system operations. With its ability to create insights and patterns from vast amounts of data, GenAI can assist in identifying areas where quantum algorithms can be applied effectively. Furthermore, GenAI can enhance the skills and capabilities of researchers and analysts in utilizing quantum computing by providing intuitive interfaces, visualizations, and decision-support tools that simplify the complexity of working with quantum systems.

Skillset Revolution through GenAI

GenAI is transforming the skill landscape across industries, democratizing complex processes, and empowering those without specialized expertise. This shift is evident in various sectors, demonstrating how GenAI is creating new opportunities for learning, **designing, and automating tasks:**

- **Language learning**

  - **Transformation**: Duolingo's GenAI customizes language learning, adapting lessons to users' progress, making advanced linguistic knowledge less necessary (Peranandam 2018).

  - **Skillset transformation**: Users engage with adaptive learning, focusing on personal progress and tailored learning paths instead of traditional language study methods.

- **Graphic design**

  - **Transformation**: Using GenAI, Canva enables users without graphic design skills to create professional designs, simplifying the process with AI recommendations (Weatherbed 2023).

  - **Skillset transformation**: Creativity and basic design intuition become the primary skills as GenAI handles technical design aspects, lowering barriers for non-designers.

- **Workflow automation**

  - **Transformation**: Zapier's integration of GenAI simplifies the automation of workflows for non-technical users, making coding knowledge less critical (Danvers, 2023).

  - **Skillset transformation:** Critical thinking and process optimization emerge as essential skills as GenAI takes over the technical execution of automation.

- **Healthcare**

  - **Transformation**: In medical imaging, GenAI aids radiologists by quickly and accurately analyzing images, enhancing diagnostic processes (Patel et al. 2019).

  - **Skillset transformation**: While radiologists' expertise remains invaluable, GenAI's assistance allows them to focus on more complex cases and diagnostics, effectively expanding their capacity to handle higher volumes of work.

Each of these examples illustrates GenAI's broad impact on traditional skill requirements. GenAI is paving the way for a future where technological tools are more accessible and inclusive, fostering innovation and creativity across various fields. This transformation emphasizes adaptability, continuous learning, and embracing new technologies to remain competitive and innovative in the evolving digital landscape.

Pillar 4: Knowledge Democratizes AI

The AI revolution is as much about the people driving it as technological advancements. Over the past decade, there has been an unparalleled democratization of AI knowledge and skills, which has played a pivotal role in its swift adoption and innovation.

- **Online learning**: Platforms like Coursera, Udacity, and edX have democratized AI education, making it accessible to anyone with an Internet connection. Prestigious institutions like Stanford, MIT, and

Harvard offer courses on pivotal AI topics, allowing learners worldwide to gain expertise without geographical or financial constraints.

- **Open-source movement**: This community has been a cornerstone in AI's evolution, significantly lowering the barriers to entry. Frameworks such as TensorFlow, PyTorch, and Keras are freely available. This accessibility empowers developers, researchers, and enthusiasts from all backgrounds to build, experiment, and innovate without the financial burdens typically associated with proprietary software. The spirit of collaboration inherent in the open-source ecosystem ensures that AI breakthroughs are widely shared, further propelling the pace of advancements and democratizing the field of artificial intelligence.

- **Collaborations and competitions**: Platforms like Kaggle have gamified AI, hosting competitions where data scientists worldwide tackle real-world challenges. These contests not only spur innovation but also serve as invaluable learning platforms. Furthermore, collaborations between academia and industry have fostered a symbiotic relationship, bridging the gap between theoretical research and its practical applications.

- **Global AI conferences and workshops**: NeurIPS, ICML, and AAAI have become hubs for AI innovation, where the brightest minds converge to exchange ideas, collaborate, and share their latest research. These gatherings foster community, networking, and knowledge dissemination.

- **Implications and responsibilities**: As AI knowledge permeates various sectors, industries worldwide integrate AI into their core operations. This widespread adoption underscores the importance of ethical considerations. As we empower more individuals with AI capabilities, it is paramount to instill a sense of responsibility, emphasizing transparency, understanding biases, and ensuring AI is harnessed for the collective good.

Navigating the Challenges of AI

While AI's promise is undeniable, it is essential to approach its potential with a balanced perspective. Like any transformative technology, AI comes with challenges and limitations. Understanding these is crucial for business leaders to make informed decisions and set realistic expectations.

- **Data dependency**: AI, particularly machine learning, thrives on data. The quality and quantity of data directly impact the performance of AI models. However, obtaining clean, unbiased, and representative data is often challenging.

- **Ethical concerns**: AI's decision-making processes sometimes lack transparency, leading to concerns about fairness and bias. Ensuring that AI models are ethical and unbiased is a pressing challenge, mainly in critical areas like healthcare, finance, and law enforcement.

- **Computational costs**: Advanced AI models, especially in deep learning, require significant computational power. This can lead to high costs and environmental concerns due to the energy consumption of massive data centers.

- **Generalization versus specialization**: While some AI models excel in specific tasks, they might not generalize to other scenarios. Building AI systems that can adapt to various situations remains a challenge.

- **Security concerns**: As AI systems become more integrated into critical infrastructures, they become cyberattack targets. Ensuring their security and robustness is paramount.

- **Human-AI interaction**: Designing AI systems that seamlessly interact with humans and understand nuance, emotion, and context is a complex challenge. While strides are being made in natural language processing and emotional AI, there is still a long way to go.

- **Regulatory and legal challenges**: As AI permeates various sectors, it brings forth regulatory and legal challenges. From data privacy concerns to accountability in AI decision-making, regulatory bodies worldwide are grappling with creating frameworks that ensure safe and ethical AI use.

- **Overcoming challenges**: The AI community is acutely aware of these challenges. Collaborative efforts are underway between academia, industry, and policymakers to address them. From developing more energy-efficient AI models to creating frameworks for ethical AI, this journey is as exciting as the potential of AI itself.

One of the most significant realizations about the importance of keeping humans in the loop came during developing a compliance platform. The project aimed to automate the compliance process, which relies heavily on standards, rules, and regulations. Initially, we believed that by simply codifying these regulations into the system, we could create a robust and effective solution.

However, as we progressed, it became clear that the existing rules could only neatly capture some compliance requirements. Numerous case-specific requirements could not be found in any written regulations. These nuances often came from the deep expertise of knowledge and domain experts—insights that were, quite frankly, only in their heads.

I remember a critical discussion with a seasoned compliance officer who casually mentioned a unique, undocumented approach to handling a specific scenario. This was pivotal—it became clear that our platform could not just be a static repository of rules and regulations but a dynamic system, integrating the expertise of professionals who had navigated real-world complexities for years. That single conversation illuminated the necessity of evolving our system to capture the unwritten practices beyond formal regulations.

To address this challenge, we designed the platform to include mechanisms for experts to input knowledge directly into the system. We also incorporated a feedback loop where these insights could continuously refine and improve the platform's recommendations. This approach ensured that while the platform automated the standard processes, it remained flexible and responsive to the nuanced expertise only humans could provide.

Conclusion

The transformative power of AI is undeniable. Its rapid evolution and adoption have been driven by a convergence of disruptions, including abundant data, unparalleled computing power, breakthroughs in machine learning and deep learning, and the democratization of AI knowledge and skills.

However, as with any powerful tool, AI's implications are vast and varied. Its potential to revolutionize industries, from healthcare and finance to entertainment and agriculture, is matched only by the ethical and societal challenges it presents. As AI becomes integral to our daily lives, businesses, policymakers, educators, and individuals must understand its capabilities and limitations.

The democratization of AI knowledge means more people than ever have the tools to harness its power. However, with this capability comes responsibility. Ensuring that AI is used ethically, transparently, and for the benefit of all will be one of the defining challenges of our era. As we continue to navigate the AI landscape, it is our responsibility to ensure that its potential is realized in a way that benefits humanity.

AI's roots have grown deep into the soil of our collective knowledge and technology, sprouting innovations that now touch every aspect of our lives. A technological phenomenon and a catalyst for broader change, it is a digital reflection of the human desire to expand our horizons.

CHAPTER 5

Disrupting Technologies and the Urge for Transformation

In a world marked by relentless change, the essence of business success has evolved beyond mere survival and hinges on transformation—profound, systemic changes that redefine an organization's trajectory. However, what exactly does transformation entail in a business context, and why does it matter? This chapter delves into the core of transformation, positing growth as a metric and the primary key performance indicator (KPI) for businesses from which all other KPIs are derived.

Transformation in business transcends the simple adoption of new technologies or methodologies. It is the gap between knowing and mastering, a leap from learning to becoming. Accurate transformation results in a permanent shift, a transcendence to a new level of mastery and capability. It is about embedding growth as the central ethos and understanding all other achievements, which are byproducts of this singular focus.

One of the most significant projects I worked on involved the integration of generative AI to automate the compliance process in construction projects. This was a pivotal moment for my client and the organization's overall direction and approach to work. GenAI proved to be a truly disruptive technology in this context, allowing us to build a proof of concept with a small, agile team in just a few weeks—something that would have been impossible using traditional methods.

The biggest challenge we faced was access to data. Compliance may seem straightforward—adhering to rules, laws, and regulations—but it is much more complex. Seasoned professionals possess countless best practices, expert knowledge, and nuanced understanding. Capturing and integrating this tacit knowledge into our AI system was essential to creating an efficient and effective solution.

We ensured that humans remained in the loop throughout the process, allowing the AI to learn from their expertise and continuously improve its recommendations. This approach automated the routine aspects of compliance and elevated the quality of decision-making by combining AI's speed and scalability with human insight. This project exemplified how generative AI, as a disruptive technology, can rapidly transform an organization's capabilities and set it on a new course of innovation and efficiency.

Reflect on the foundational principles laid out in this book's introduction. A continuous quest for learning and growth fuels the journey of both personal and organizational transformation. Organizations must foster a culture of curiosity and constant improvement to stay ahead in the rapidly evolving digital arena.

Drawing parallels from the human experience, growth and transformation manifest through two primary paths: *kensho* and *satori*. *Kensho* signifies growth through pain—a reactive transformation triggered by adversity. *Satori*, on the other hand, represents growth through insight—a proactive, perception-driven evolution. While *kensho* might be more common, offering valuable lessons in resilience, *satori* presents an aspirational, transformational model through deliberate, conscious insight.

In business, the journey from *kensho* (growth through adversity) to *satori* (growth through insight) mirrors the transition from reactive to proactive transformation. Imagine a company that initially scrambles to digitize operations in response to a sudden market shift (akin to *kensho*). Over time, this company could evolve to adopt a *satori* approach that strategically leverages data and AI—not just for immediate needs but as part of a visionary strategy for sustained innovation. This evolution exemplifies the shift from merely surviving in a disruptive environment to thriving through foresight and insight. This illustrates how businesses can embody these spiritual principles to navigate the tumultuous waters of the digital age.

We are navigating an epoch of disruption reminiscent of the early days of the Internet revolution. Businesses face a "sink or swim" scenario, propelled by the tidal forces of digitalization, data proliferation, and AI. The continuous upheaval in customer behavior, economic landscapes, and global challenges like pandemics only accentuate this reality. The pertinent question is whether businesses can pivot from merely reacting to such disruptions to proactively driving growth through insight.

Data, AI, and technology serve as powerful enablers in this journey. However, their potential cannot be fully unleashed when seen only as solutions to immediate problems rather than being integrated into a broader strategy for insight-driven growth. These tools offer the means to transcend traditional limitations, allowing businesses to anticipate changes, innovate proactively, harness disrupting technologies, and understand changes as opportunities for evolution.

Moving a business towards growth through insight requires a paradigm shift—from viewing transformation as an episodic response to external pressures to embracing it as an ongoing strategic imperative. This involves cultivating a culture where continuous learning, agility, and adaptability are paramount, one in which data and AI are not just operational tools but foundational elements of strategic foresight and innovation.

Businesses must remain agile, ready to adapt their models and strategies in response to emerging data trends and market dynamics. This adaptability ensures business resilience and paves the way for transformative success.

A business, much like an individual, has the potential to achieve its true essence through transformation. This does not imply a linear path but a dynamic, iterative process of constantly aligning with core purposes, adapting strategies, and embracing new engagement, operation, and value creation models. Transformation, therefore, is both the journey and the destination—a continual process of becoming those shapes not just what a business does but fundamentally what it is.

In the transformation era, insight guides businesses toward their true potential, mirroring the human quest for growth (through *kensho* or *satori*) and painting a vivid picture of the possibilities beyond the horizon of known challenges. Clarity of purpose, strategic insight, and judicious use of data and AI will delineate those who merely survive from those who genuinely transform in the face of disruptive forces.

This chapter lays the foundation for understanding transformation not just as a necessity but as the most profound opportunity for businesses to realize their fullest potential by achieving growth that is both sustainable and deeply aligned with their core mission. AI's projected economic impact and strategic importance extend beyond disruption and are double-edged swords, presenting business challenges and opportunities. Transitioning from understanding the "why" of transformation to navigating the "how" will mean realizing that data and AI can spark a profound, insight-driven business evolution.

The Market Value of AI

The AI market is projected to reach a staggering USD 1,581.70 billion by 2030, driving unprecedented growth across industries. Companies must harness AI's power and embed it into their strategic planning and execution (MarketsandMarkets n.d.).

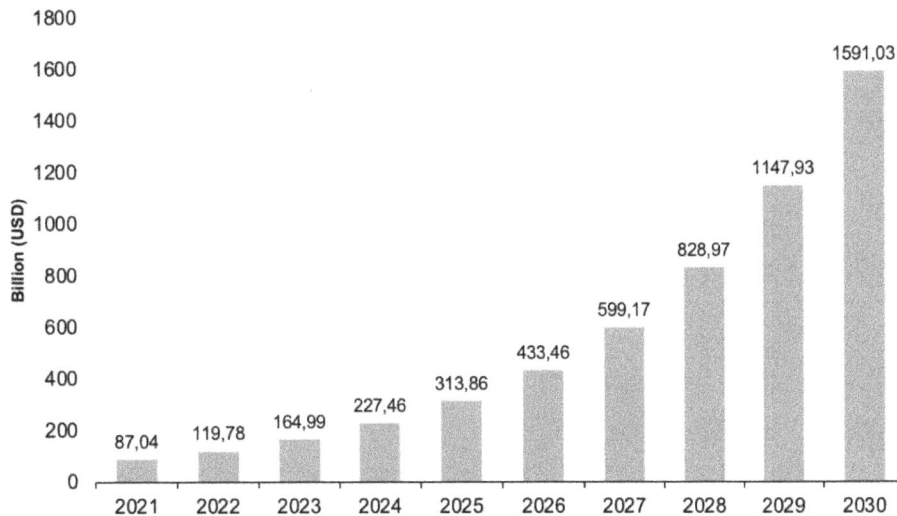

Figure 19. AI Market Size by 2030 **(MarketsandMarkets n.d.)**

While the projected market size reflects AI's growth trajectory, it is essential to note that achieving this potential requires addressing challenges and considerations such as data privacy, ethical use of AI, regulatory frameworks, and the need for skilled AI talent. Organizations and policymakers must work together to create an ecosystem that fosters responsible and ethical AI adoption while reaping the benefits of this transformative technology. Balancing optimism by addressing pressing ethical considerations is prudent as generative AI's possibilities are embraced.

Transformation: Rethinking the Business for Sustained Success

In today's business landscape, disruptions are more frequent and intense. They are significant changes or shifts that challenge existing norms, business models, and industry dynamics. Technological advancements, changes in consumer behavior, market trends, or external forces can trigger them. Each disruptive wave brings new opportunities and threats that demand a strategic response from businesses.

Organizations must embrace transformation, which goes beyond mere adaptation or quick fixes—it involves a fundamental rethinking of the entire business to ensure long-term viability and sustained success. This requires a shift in mindset, strategy, processes, and culture to align with the market's new realities. Transformation is about envisioning and creating a new future for the business rather than clinging to outdated practices or relying on temporary solutions.

A common misconception is that transformation can be achieved by adopting a few digital technologies or making superficial changes. However, true transformation requires a holistic approach that permeates every aspect of the organization, leveraging technology, data, and innovation to reimagine the value proposition, business models, customer experiences, and operational processes.

To illustrate this, consider the emergence of direct-to-consumer (DTC) brands. Traditional retail faced significant challenges with the rise of e-commerce, compounded by changing consumer behaviors during the COVID-19 pandemic. DTC brands like Warby Parker (eyewear) and Glossier (cosmetics) have successfully bypassed traditional retail channels, leveraging social media and digital marketing to build direct consumer relationships. This approach allows for personalized experiences and rapid market traction, forcing established retailers to rethink strategies and accelerate digital transformation efforts (Joshi 2020).

Such shifts demonstrate the urgency for traditional businesses to reinvent themselves to remain competitive in the face of new, digitally native entrants. By embracing digital technologies and direct engagement with consumers, even legacy brands can find pathways to innovate and grow in the digital age.

These examples illustrate the necessity to reimagine businesses in response to disruption, highlighting the importance of being proactive, agile, and open to new ways of operating and delivering value. Transformation involves breaking free from legacy mindsets.

It is crucial for organizations not just to adapt but actively challenge prevailing norms and assumptions. For instance, Zoom revolutionized the communication industry by providing a reliable, user-friendly platform for video conferencing, which became essential during the COVID-19 pandemic (Qumer & Ikrama 2021). Similarly, Robinhood disrupted the financial sector by democratizing stock trading, offering commission-free trades and user-friendly app interfaces that appealed to a new generation of investors. On the other hand, Shopify has been a significant force in the transition toward digital commerce, pioneering user-friendly e-commerce platforms that enable businesses of all sizes to sell online effectively (Whitehorn, n.d.; Nitro Logistics, 2023). These companies exemplify how embracing disruptive technologies and models can position businesses for long-term success and resilience, even as they navigate the challenges posed by rapidly shifting market landscapes.

Learning from Failures

MySpace, Yahoo, **and BlackBerry once dominated their respective markets—but then faced challenges that led to their decline:**

- **MySpace**: Once a dominant social networking platform before the rise of Facebook. While MySpace initially gained popularity, it needed to invest in building its robust platform and expanding its capabilities. Instead, it relied on existing infrastructure and partnerships, limiting its ability to adapt to changing user preferences and technological advancements. As a result, its competitive edge was lost, and soon, it was overshadowed by Facebook's comprehensive platform, which offered a wider range of features and integration opportunities.

- **Yahoo**: Yahoo was once a leading Internet portal, offering various services such as search, email, news, and more. However, needing to transform and compete effectively with companies like Google, Yahoo relied on partnerships and outsourcing critical services rather than investing in building the platform and expanding capabilities. This limited its ability to innovate and respond to evolving user needs. Ultimately, Yahoo's failure to fully transform and build a robust platform led to its decline and acquisition by Verizon.

- **BlackBerry**: BlackBerry was a pioneer in the smartphone market, known for its secure messaging and email capabilities. However, it failed to adapt when competitors like Apple and Google introduced more advanced smartphones and app ecosystems. Instead of investing, BlackBerry relied on its strengths in secure messaging and enterprise services. This limited its ability to compete in the rapidly evolving smartphone market.

These examples illustrate the importance of building comprehensive platforms and leveraging additional data and AI transformation capabilities. Relying solely on platform-as-a-service without investing in a bespoke platform hinders an organization's ability to innovate, respond to market changes, and fully harness the potential of data and AI. Successful transformation requires adopting technology and building a solid foundational ecosystem that enables organizations to leverage data, AI, and additional capabilities to stay ahead of the competition.

Another example involves organizations that viewed data and AI as buzzwords rather than catalysts for holistic transformation. These companies only scratched the surface, believing that superficial changes—like incorporating AI algorithms into existing processes or implementing basic data analytics—would suffice. They should have recognized that true transformation required completely reimagining their business models, structures, and operations.

Such organizations took their existing business practices and plastered them onto the digital landscape, failing to grasp the profound shifts needed to thrive in the digital age. As a result, they needed help keeping up with the pace of innovation and were eventually overshadowed by competitors who embraced full-scale data and AI transformation. **Here are a few instances highlighting the challenges and pitfalls of failing to adopt or correctly implement technologies fully:**

- **GE's digital transformation**: General Electric's ambitious attempt at a digital transformation was characterized by missteps, including a lack of clear definitions, phased development, and quantifiable KPIs to measure success. The endeavor suffered from insufficient buy-in at various levels and lacked a dedicated management team driving the new initiative. This case emphasizes the importance of clear vision, measurable goals, phased implementation, and strong leadership in digital transformation projects (Rohn 2022).

- **Microsoft's AI editorial failed.** Microsoft's decision to replace human editors with AI to manage news articles on its MSN platform led to several issues, including publishing fake news and inappropriate content. This failure underlines the critical need for thorough AI model development, ongoing monitoring, and a balanced human-AI collaboration to manage risks and maintain quality control in content curation (Rohn 2022).

- **Work shifting to remote during COVID-19**: While not a failure per se, the rapid pandemic-induced shift to remote work highlighted several challenges, including the risk of reverting to old habits, the necessity of continuous IT investment, and the importance of maintaining a clear digital strategy beyond emergency measures. Companies learned that changes driven by necessity during the crisis could serve as valuable lessons for long-term digital strategy, emphasizing the need for a sustained approach to digital transformation (7T n.d.).

Such cautionary examples remind us that true transformation demands more than quick wins or surface-level changes. Instead, it requires a deep commitment to building a solid foundation, leveraging data and AI as strategic assets, and rethinking every aspect of the business.

Just as mountaineers must endure arduous climbs to reach the summit, companies must embrace the transformation journey and take small but purposeful steps forward. There is no one-size-fits-all approach or shortcut to success. Each company must embark on its unique journey—learning, adapting, and evolving. By doing so, they can harness the full potential of data and AI to scale new heights and secure a competitive advantage in the ever-changing business landscape.

Success Stories from the Field

- **Lemonade**: A disruptor in the insurance industry, Lemonade leveraged AI to simplify the consumer process. Their chatbot, Maya, automates policy quotes and claims, making insurance more accessible and affordable. Its AI-driven model has also been integral in fraud detection, contributing to swift claim resolutions—sometimes in seconds. This innovative approach has spurred significant growth for the company and reshaped industry standards (Klein, 2023).

- **Stitch Fix**: Its AI-powered personal styling service has transformed the retail clothing industry. By combining data analysis with the expertise of personal stylists, it offers customized clothing selections that align with each customer's style and fit preferences. This unique fusion of AI with human judgment has led to higher customer satisfaction and reduced returns. It has driven exponential growth for the company, proving the value of personalization in retail (Davenport 2021).

- **Duolingo**: This AI-enhanced platform has made language learning widely accessible and engaging by personalizing educational content to match users' learning styles and paces, which provides a customized experience that traditional language programs struggle to offer. With its vast user base and high engagement levels, Duolingo demonstrates the impact of AI in education and has transformed how millions of people learn new languages (Marr, 2023).

Each of these companies has utilized AI to improve existing processes and lead the way in their respective fields, creating new value for customers and redefining market expectations.

Ethical Considerations

While celebrating the disruptive potential of generative AI, it is essential to navigate its ethical considerations. As AI systems increasingly influence our daily lives, ensuring they operate responsibly is crucial. Ensuring fairness, transparency, and accountability is crucial to avoiding specific biases, discrimination, or unintended negative consequences.

To combat such challenges, organizations must establish robust ethical frameworks and guidelines for the responsible development and deployment of AI systems. This commitment involves proactive measures safeguarding privacy, enhancing security, and ensuring AI's positive social impact.

For example, one unintended consequence of AI is algorithmic bias, where AI systems inadvertently perpetuate societal biases. A well-known case is the gender bias observed in AI recruitment tools that favored male candidates due to biased training datasets. To address this, companies have had to improve data diversity and implement algorithmic audits to detect and correct biases.

Another example is the impact of facial recognition technology on privacy and rights. Concerns have led to companies like IBM stepping back from offering general-purpose facial recognition services to law enforcement due to potential misuse associated with racial profiling and mass surveillance.

In navigating rapid technological changes, organizations must adapt their business models and operations to integrate AI ethically. By doing so, they can safely and transparently leverage AI to transform their business models and maximize market impact.

Harnessing AI's transformative potential must be anchored by an ethical framework mirroring the broader principles of growth and transformation that are part of our personal and spiritual journeys. Such alignment not only fosters technological innovation that is fair, transparent, and accountable but also ensures that advancements contribute positively to the collective well-being, echoing the spiritual principle that actions and creations should harmonize with and uplift society's collective consciousness.

Here are a few examples of ethical dilemmas that plagued the rollout of AI systems:

- **Addressing gender bias:** During its pilot test, Amazon's AI recruiting tool exhibited gender bias, favoring male candidates for technical roles. This raised significant ethical concerns (Andrews & Bucher, 2022).

  - **Discovery**: Researchers and analysts noticed discrepancies in the tool's recommendations, prompting further investigation.

  - **Origins**: The gender bias stemmed from the historical data used to train the algorithm, primarily from resumes submitted to Amazon over a decade, which were predominantly from male candidates due to existing gender imbalances in the tech industry.

- **Tackling discrimination**: Apple Card faced accusations of gender discrimination in its credit limit algorithm, with women receiving lower credit limits than men with similar financial backgrounds (CBS News 2019).

- **Discovery**: Accusations surfaced when users reported disparities in credit limits between men and women, prompting media scrutiny and public attention.

- **Origins**: The disparity was likely due to inherent biases in the algorithm or the data it was trained on, reflecting broader societal inequalities in financial systems.

- **Fake news and biased algorithms**: Facebook has grappled with challenges like the proliferation of fake news and potential biases in its algorithms (Pickup 2021; Menczer 2021).

  - **Discovery**: Concerns arose from increasing user complaints, media scrutiny, and academic research highlighting the platform's role in spreading misinformation and promoting polarizing content.

  - **Origins**: Issues originated from Facebook's algorithmic design, which prioritized user engagement and viral content without adequately addressing the spread of misinformation or the amplification of biased narratives.

- **Facial recognition**: Like other tech giants, Microsoft's facial recognition tools were not 100 percent accurate in recognizing individuals with dark skin (Najibi 2020).

  - **Discovery**: Microsoft's decision not to provide facial recognition services to law enforcement agencies was prompted by concerns over racial discrimination and privacy violations raised by civil rights groups, activists, and internal discussions within the company.

  - **Origins**: Concerns stemmed from growing public awareness of the potential for misuse, racial profiling, and privacy infringements associated with facial recognition systems that law enforcement agencies were deploying.

- **Perspectives on general-purpose AI**: IBM has taken a firm stance against working on general-purpose AI due to the potential risks, while OpenAI believes its development is inevitable (IBM, n.d.; OpenAI, n.d.).

  - **Discovery**: The clash between IBM and OpenAI arose from divergent approaches and philosophies within the AI research community. IBM advocates for narrow AI solutions tailored to specific tasks, while OpenAI pursues broader research goals.

  - **Origins**: The clash reflects differing opinions on the risks and benefits of general-purpose AI and competing priorities and strategies among industry players and research organizations.

Ethical implications cannot be ignored as AI continues to permeate various sectors. Companies must proactively address concerns, ensuring AI systems are transparent, fair, and beneficial for all users. Learning from real-world examples can pave the way for responsible AI development that respects human rights and societal values.

Navigating the ethical landscape of AI development and application is reminiscent of deeper, more profound journeys to align with our technological aspirations and spiritual and moral compass. Ethical considerations pave the way for a transformative approach transcending mere compliance with guidelines, but instead envisioning AI and data analytics acting as instruments for positive change that reflect our highest values and most profound understanding of interconnectivity.

This perspective is not new. It echoes timeless wisdom about the interconnectedness of all life and the importance of aligning our actions with universal principles of harmony, respect, and empathy. Just as spiritual journeys involve deep reflection, learning, and growth, so too does the path of AI transformation, which will require us to question, learn, and evolve—technically, ethically, and spiritually. By embedding these values into the fabric of technological endeavors, a foundation for AI can be created that enhances efficiency and innovation and uplifts and enriches the collective human experience.

The Spiritual Parallels of AI Transformation

AI transformation transcends mere technical upgrades or efficiency enhancements. It invites infusing technological endeavors with mindfulness and empathy, principles reminiscent of spiritual wisdom. Such an approach is consistent with the rigor and ambition behind AI innovations, enriching them with a vision for technology that uplifts humanity and fosters a more empathetic and understanding world.

The path of transformation—whether personal, organizational, or technological—is not just about achieving external milestones but also about aligning with a purpose that transcends immediate gains. It is about embedding a sense of responsibility towards the broader ecosystem in which we operate, ensuring that technological advances contribute positively to the well-being of all stakeholders. It is a paradigm of growth that harmonizes technological prowess with ethical integrity and spiritual insights that must be committed to. **The AI and personal/spiritual transformations share foundational principles:**

- **Self-awareness and reflection**: Just as spiritual seekers embark on introspective journeys to understand their inner selves, organizations must reflect upon their technological capabilities, recognizing strengths and areas for improvement.

- **Guidance and mentorship**: Guidance from mentors or sacred texts is invaluable in spiritual quests. Similarly, expert guidance and industry benchmarks can illuminate the path forward in AI.

- **Continuous learning and adaptability**: Spiritual growth is an ever-evolving process that demands adaptability and constant learning. Similarly, the dynamic world of AI necessitates an unwavering commitment to education and evolution.

- **Ethical considerations**: Morality is central to many spiritual paths. Ethical considerations like fairness and transparency are optional and essential in AI.

- **Overcoming challenges with resilience**: Spiritual seekers often confront profound challenges, requiring deep resilience. As organizations navigate the AI journey, they will face obstacles, underscoring the importance of perseverance.

- **Holistic approach**: True spiritual enlightenment embraces the mind, body, and soul. Similarly, a successful AI transformation is holistic, encompassing not just technology but also culture, ethics, and business strategy.

- **Purpose and vision**: A clear purpose often propels spiritual journeys. For organizations, lucid ideation ensures that technology serves a larger mission and aligns with overarching goals.

- **Community and collaboration**: Spiritual paths often emphasize collective growth and community strength. In the AI domain, collaboration can be a catalyst, fostering innovation and shared learning. In data and AI, encouraging knowledge sharing and valuing feedback is crucial to refining algorithms and enhancing model accuracy. My experience navigating the startup ecosystem underscores the indispensable value of collaborative efforts in driving innovation.

Figure 20. Transformation with Insight

Drawing from these spiritual parallels, we are reminded that AI transformation is not just a technical endeavor but a profoundly human one. Facing a future where AI augments our capabilities and reflects our deepest values, we find ourselves at a crossroads that is as much about spiritual awakening as technological advancement. This confluence of paths underscores a shared journey of transformation transcending boundaries between the personal and the collective, between the material and the spiritual. Embracing this journey requires leveraging AI for ethical and sustainable growth and imbuing endeavors with a consciousness that honors our interconnectedness with all of existence. This approach resonates with the wisdom that every step towards technological innovation can achieve greater harmony and understanding, mirroring our transformative journeys toward personal growth and enlightenment.

Conclusion

The journey ahead requires us to react to the unfolding landscape of AI and data analytics and proactively shape it to reflect our collective aspirations for a world where technology serves the highest good. The ethical considerations and spiritual parallels associated with AI transformation challenge us to reimagine innovation and envision future scenarios where AI amplifies our capabilities while aligning with principles of fairness, transparency, and the greater welfare of all. In the nexus of anticipation and ethical foresight, businesses can forge strategies to lead with integrity, ensuring their AI initiatives contribute to a just and thriving society.

By embracing a culture of continuous learning, ethical commitment, and strategic foresight, businesses can navigate the future with confidence and purpose, ready to capitalize on new opportunities presented by AI and data analytics.

As we stand on the brink of a new era defined by AI, it is crucial to recognize that this transformation is not merely a technological leap but an opportunity to align business practices with deeper values of connection, empathy, and sustainability. This AI journey, mirroring personal and spiritual growth, challenges us to embrace technological advances with a consciousness that uplifts humanity and the planet. While AI's market value is essential, let us carry forward the wisdom that true transformation integrates innovation with an ethical and spiritually conscious approach, promising a future where technology and human values coalesce for global well-being and prosperity.

CHAPTER 6

Transformation Engineering: The Wisdom Spire Framework

Adopting a reality model for an organization/business revolves around harnessing the power of data, AI, and GenAI to build robust insight capabilities. This concept, inspired by Vishen Lakhiani's *The Code of The Extraordinary Mind*, parallels how individuals create models of reality in their minds based on belief systems, perceptions, and projections. Similarly, businesses can develop a reality model to navigate their environment effectively.

In this context, a business's reality model involves recognizing the immense value of gathering and amalgamating vast amounts of external and internal data to comprehensively understand customers, market trends, and business dynamics. This model helps businesses interpret complex data, predict future trends, and make informed decisions.

McKinsey & Company highlights the importance of such models in adapting to technological advancements and changing market conditions, advocating for continuous learning and innovation across all organizational levels (Van Kuiken, 2022).

By leveraging advanced technologies, organizations can construct a dynamic picture of their operational landscape, enabling them to remain agile, proactive, and strategically aligned with their goals. This is a recognition of the immense value of gathering vast amounts of external and internal data and amalgamating them to understand customers, market trends, and business dynamics comprehensively.

Leveraging AI and GenAI capabilities means efficiently processing and analyzing this diverse data landscape. Advanced algorithms and machine learning models extract actionable insights.

The ultimate goal is to empower everyone with personalized insights directly relevant to their tasks and responsibilities. Providing the right insights at the right time enables employees to make data-driven decisions, enhance productivity, and drive meaningful outcomes.

This gives an organization/business a significant competitive advantage by more deeply understanding customer preferences, behaviors, and needs. It allows for the tailoring of products and services that exceed their expectations. The untapped market potential can be uncovered, and emerging trends that identify new opportunities and innovative solutions can be realized. Moreover, data-driven insights enable the optimization of operations, improved efficiency, and more informed strategic decisions.

To support this model of reality, developing and integrating advanced data analytics tools, AI technologies, and GenAI capabilities are prioritized. Investments in robust infrastructure are required to ensure high-quality data collection, storage, and processing. Furthermore, a culture that values data literacy, experimentation, and data-driven decision-making must be cultivated throughout the organization.

Every transformational journey, personal or organizational, begins with a deep dive into the foundational beliefs and the models that shape our reality. These models, constructed from collective experiences, beliefs, and perspectives, serve as the bedrock for actions, decisions, and strategies. Businesses are shaped by their unique models of reality, which dictate operational frameworks and decision-making processes.

Recognizing and reassessing such models is crucial as it allows the alignment of strategies within an evolving landscape, ensuring that the transformation journey is purposeful and impactful. The shared path of evolution, learning, and adaptation is underscored by drawing parallels between such journeys by individuals and those embarked upon by organizations. This convergence of personal and organizational transformation illuminates the intrinsic link between the growth of individuals and the progressive evolution of businesses in navigating the complexities of the modern world.

This chapter ventures beyond merely upgrading models of reality to conscious engineering and the profound impacts of data and AI into the fabric of organizational transformation. Conscious engineering catalyzes growth and innovation by facilitating a deeper understanding and application of data and AI. By aligning these technological advancements with core models of reality, a comprehensive journey that spans redefining foundational beliefs to the harnessing of the full potential of AI is embarked upon. This conscious approach to technological integration illustrates how the components—models of reality, conscious engineering, and the strategic application of AI—are intrinsically linked in driving forward growth and innovation.

Models of Reality in Organizations

In business, reality models are deeply influenced by historical data, past experiences, and established processes. They serve as the blueprint for strategies, decision-making, and interactions with the world. However, as technological advancements accelerate and consumer behaviors shift, it becomes imperative to reassess and update foundational models continually. Adhering to outdated ones can hinder innovation and growth while adapting them in response to new insights and trends can unlock unprecedented opportunities for advancement.

Understanding reality informs every aspect of organizational life, from the systems used for daily operations to the strategic initiatives pursued for long-term growth. In an organization or business context, adopting a reality model centered around integrating data, AI, and GenAI enables the forging of a powerful insight-driven approach that values extensive data collection, analysis, and harnessing advanced technological capabilities to distill actionable insights.

The aim is to democratize access to such insights, ensuring that every member of an organization is empowered with information tailored to their specific roles and responsibilities. This democratization is not merely about providing access but enabling informed decision-making, fostering productivity, and driving meaningful innovation at every level.

This dynamic and insightful model of reality lays the groundwork for a transformative shift in organizational mindset—a shift towards conscious engineering. It is crucial to recognize that upgrading an organization's mindset through conscious engineering is not just an operational necessity but a strategic imperative that aligns with an evolving understanding of reality. By actively questioning, challenging, and reshaping models, the business landscape can be adapted to and shaped, paving the way for a future where the organization thrives on innovation, inclusivity, and sustainable growth.

Conscious Engineering: Upgrading the Organizational Mindset

Conscious engineering empowers the recognition that models of reality are not fixed and can be changed, including actively and intentionally redefining an organization's model of reality. It is a concept derived from an interdisciplinary approach combining psychology, organizational behavior, and technology management.

The framework encourages organizations to critically evaluate and adapt their core beliefs, processes, and strategies. Originating from the need to navigate the complexities of modern business environments, conscious engineering emphasizes agility, innovation, and ethical considerations. By challenging existing beliefs, questioning established processes, and embracing new ways of thinking, organizations can align their models of reality with the current business landscape and future aspirations. Applied across various sectors, it proves vital for organizations aiming to remain competitive and responsive to change.

For organizations, the journey toward becoming data-driven and leveraging AI begins with acquiring data that reflects current reality. Data-driven understanding enables organizations to redefine business processes based on concrete information and move away from outdated assumptions. **Conscious engineering involves:**

- **Data acquisition**: Gather relevant information from various sources to comprehensively understand the business environment.

- **Data analysis**: Using advanced tools and techniques to understand and extract meaningful insights from the data deeply.

- **Insight implementation**: Applying the understanding gained from data analysis to redefine business processes, strategies, and decision-making frameworks.

The Role of AI in Conscious Engineering

AI plays a pivotal role in conscious engineering. Its ability to process vast amounts of data, recognize patterns, and generate insights provides a deeper understanding of current reality and potential opportunities.

The first tranche of required data involves gaining visibility and understanding of the organization, including unlocking valuable information about operations, customers, and markets through business intelligence (BI) dashboards, data-driven reports, and insightful analyses. For example, a retail company may use a dashboard to track real-time sales data, customer feedback, and inventory levels, enabling data-driven decisions regarding pricing, inventory management, and marketing campaigns.

The role of AI, particularly GenAI, can be pivotal in this process. GenAI encompasses advanced AI technologies capable of analyzing vast amounts of data, uncovering hidden patterns, and generating powerful insights. With GenAI, gaining visibility and understanding becomes faster and more accessible to all organization members rather than being the domain of data scientists alone. For instance, customer service representatives can utilize AI-powered chatbots to swiftly access customer information, previous interactions, and sentiment analysis, enabling them to provide personalized and efficient customer support.

In an era when change is the only constant, we find ourselves at a pivotal moment reminiscent of the Netscape browser's impact on the Internet. This comparison, first highlighted by the *New York Times* (Metz

2023), is a metaphor for the current transformation of GenAI, including tools like ChatGPT and Gemini (formerly Bard). Just as Netscape illuminated the World Wide Web's untapped potential, GenAI is ushering the once arcane discipline of artificial intelligence into the public consciousness, marking the dawn of an unprecedented era of technological empowerment and disruption.

GenAI is set to reshape the technological landscape, particularly its advanced forms. This includes democratizing access to AI tools, fostering innovation, accelerating developments in data management, computational power, and skill enhancement. GenAI enables the creation of new, synthesized data, bringing insights from dark data to light and ensuring the prioritization of critical data elements, thereby automating and improving data quality.

What is GenAI? Advanced algorithms that create or generate new content, solutions, or data that mimic real-world patterns and insights. It leverages vast datasets to produce outputs ranging from text, images, and music to complex predictive models. By doing so, GenAI is not just automating tasks but is also fostering creativity, enhancing decision-making, and paving the way for innovative solutions across industries. Applications span the automation of content creation to revolutionizing product design. It is poised to reshape the technological landscape, democratizing access to AI tools and fostering innovation by tapping into diverse perspectives. This broad access to AI tools will allow individuals and organizations to harness AI's transformative power.

Diving into GenAI's potential, **it is crucial to recognize its transformative implications:**

- **Fostering creativity:** GenAI offers a fresh canvas for professionals across industries. For instance, it is being used in the entertainment industry to develop new scripts and plots, pushing the boundaries of storytelling. In fashion, designers leverage it to create innovative patterns and styles, redefining trends. Meanwhile, music sees AI composing pieces that blend classical influences with modern sounds, inviting artists and musicians into uncharted creative domains. This technology opens new possibilities, inspiring professionals to explore beyond conventional limits.

- **Enhancing task performance**: GenAI's adaptability enables undertaking new tasks with minimal training, drastically improving efficiency across industries. It automates complex design processes in engineering, accelerates diagnostic accuracy in healthcare, and crafts personalized customer service responses. Zero-shot learning further expands GenAI's capabilities, allowing it to perform tasks without exposure to pre-training data. This includes generating legal documents or educational content in new domains and showcasing its ability to adapt and provide solutions across varied fields without extensive pre-programming.

Beyond its direct influence, GenAI also accelerates progress in pivotal areas like data management, computational power, and skill enhancement.

The Wisdom Spire Framework: Ascending the Data and AI Summit

In an age of inevitable and accelerating change, the question arises, "Can businesses proactively engineer transformation, or must they remain at the mercy of the next wave of disruption?"

This pivotal inquiry has led to my creation of the Wisdom Spire Framework, a comprehensive blueprint inspired by the principles of human transformation. It is designed to empower leaders of small and medium enterprises and captains of industry within large organizations to navigate and shape their enterprises' futures with intention and foresight.

Figure 21. The Scaffolding of Insight: From Mapping to Transformational Engineering

The framework is born from lessons learned and insights gained from the front lines of business innovation and the deep wells of human growth and spiritual evolution. It is a testament to the belief that transformation can be engineered and that organizations can set the sails of their destiny in the tumultuous seas of market disruption and technological advancement.

At its core, the Wisdom Spire Framework invites business leaders to reframe their approach to change. It challenges the reactive mindset that waits for external forces to dictate the need for transformation. Instead, it advocates for a proactive stance, readying organizations to leverage change for growth, innovation, and a lasting positive impact on the world and humanity.

This is the heart of my narrative, offering a methodology and a philosophical shift toward seeing businesses as dynamic entities capable of conscious evolution. By applying it, leaders can unlock the potential to scale their operations, transform their business models, and contribute to a ripple effect of positive change that extends far beyond the confines of their companies.

The Wisdom Spire Framework charts a course toward meaningful and sustainable transformation by weaving together the threads of data-driven insight, artificial intelligence, and the timeless wisdom of human development. To give it a grounding in the tangible and relatable, I use the metaphor of mountain climbing—

an analogy reflective of my passion for mountains and the transformative journeys they represent. Just as every ascent is a narrative of overcoming, adaptation, and revelation, so is the business transformation journey, which embodies the philosophy that each step on the path is as significant as reaching the summit.

Figure 22. Ascending the Mountain of Transformation

Transformation is more than just the tools or technology (the gear). It involves extensive preparation, strategic planning, and an understanding that no matter how meticulously one prepares, one must navigate unforeseen challenges, pivot strategies, and, at times, retreat and regroup.

The Wisdom Spire Framework emphasizes readiness for adaptation, the humility to learn from the journey, and the resilience to continue moving forward, even when the path becomes unclear or treacherous. It is about embedding a culture of continuous growth, learning, and evolution at the heart of one's business strategy—mirroring the preparations and mindset needed to conquer a mountain.

The Wisdom Spire Framework, which I developed through years of experience in leading transformation projects, is designed to guide organizations through the complexities of modern business challenges. It integrates technical, strategic, and human elements into a cohesive approach, ensuring that transformation is not just about change but about meaningful, sustainable evolution.

One of the most impactful applications of the Wisdom Spire Framework was in a project involving a major digital transformation for a multinational corporation in the manufacturing sector. The company faced significant challenges due to outdated systems, siloed data, and a rigid organizational structure stifling innovation. It struggled to compete in an increasingly digital marketplace and needed a comprehensive transformation to stay relevant. **The initial assessment revealed several deep-rooted issues:**

- **Siloed data**: The company's data was fragmented across various departments, making it challenging to gain comprehensive insights and make informed decisions.

- **Outdated systems**: Legacy systems needed to be more efficient and compatible with modern technologies, which hindered the adoption of digital tools.

- **Rigid organizational structure**: Hierarchical and resistant to change, with a risk-averse culture slow to adapt.

Given the transformation's complexity and scale, the Wisdom Spire Framework was an ideal approach. It allowed us to address these challenges systematically, ensuring that each aspect of the transformation was aligned with the overall strategic goals.

The Seven Components of the Wisdom Spire Framework

The Summit Vision

Just as mountaineers meticulously plan their route to the summit—defining their "why" before they begin—organizations embarking on data and AI transformation must also start with their ultimate vision. This strategy encapsulates the purpose behind the organization's journey to leverage data and AI for transformative success.

In his influential work *Start With Why*, Simon Sinek sheds light on the critical importance of understanding and articulating the purpose behind our actions and the existence of our organizations (Sinek 2011). This approach explores the concept of purpose and delineates why specific individuals and organizations achieve higher success and influence. Sinek introduces the Golden Circle model, emphasizing starting with "why" before addressing the "how" and "what." This model suggests that most organizations begin by defining what they do, followed by how they do it uniquely, often sidelining the essential question of why they are doing it.

This insight is pivotal. A robust data and AI strategy must begin with a clear and compelling "why." This is not merely about the technologies to be implemented or the innovative methods to be adopted but about articulating a vision and mission that propels the organization toward its transformative summit.

By anchoring in "why," organizations imbue all levels and functions with a vital purpose. This unified direction is a guiding light, aligning employees, stakeholders, and partners around a shared mission. Understanding the purpose of the contributions makes individuals more engaged, motivated, and committed to delivering their utmost effort.

Mirroring the preparation climbers undertake to define their summit vision before beginning their ascent; organizations must craft a clear and visionary data and AI strategy. This strategy should outline direction, set strategic objectives, and unite stakeholders towards a shared goal, ensuring that every step taken is aligned with and contributes to the overarching mission of transformation.

This will encourage organizations to move beyond superficial layers of technological adoption and prompt a deeper reflection on the transformation's purpose. This ensures that the journey is about achieving technological milestones and realizing a vision that resonates with and motivates all involved.

Value Tracking

In mountaineering, every decision—from selecting the right path to adjusting for weather conditions—requires meticulous planning and prioritization based on current data concerning terrain, weather patterns, or the climber's physical state. Such strategic evaluation mirrors the decision-making process in leveraging data and AI. Just as climbers assess various factors to ensure a safe and successful ascent, businesses must evaluate potential business value, technical feasibility, and alignment with strategic goals.

Value tracking is akin to a mountaineer's careful monitoring of conditions and readiness to alter plans. This system allows organizations to measure the impact of each initiative, prioritize resources effectively, and steer projects toward the highest return on investment and strategic relevance.

Technology Ecosystem

In mountaineering, gear selection is a deliberate process. Each piece is chosen for its specific utility in overcoming the mountain's unique challenges. Similarly, a technology ecosystem comprises the foundational tools for data collection, storage, analysis, and AI model development. The strategic integration of these platforms enables seamless collaboration and efficient data workflows.

Investing in the right technology ecosystem ensures that an organization can scale new heights of innovation and efficiency. By carefully selecting each component—from cloud storage solutions to advanced analytics platforms and AI development tools—businesses can create a cohesive, powerful toolkit to support strategic objectives and drive transformative outcomes.

Having the right tools and understanding how they fit together is essential to creating a comprehensive system that empowers the organization. As climbers must know how to use their gear effectively, organizations must master their technology ecosystem to unlock their full potential in the data and AI landscape.

Data Foundation

A solid base camp is fundamental for a successful mountain expedition. Likewise, creating a robust data foundation is essential to the data and AI transformation journey. This involves meticulously setting up data governance frameworks, management practices, and infrastructure for maximum efficiency, safety, and accessibility.

This is like securing a base camp against environmental hazards, ensuring it provides a reliable launchpad for the ascent. A robust data foundation encompasses comprehensive data integration, architecture, and storage solutions that will be the backbone of an organization's entire data ecosystem.

A well-structured data foundation empowers organizations to trust their data for decision-making and to scale their AI and data-driven initiatives effectively. It is about laying the groundwork that ensures data flows seamlessly, insights are derived efficiently, and strategic objectives are supported by a reliable, accessible, and secure data environment.

Talent and Operational Model

Successful mountain expeditions require carefully selected climbers with unique skills and expertise. Data and AI transformation demands a talented workforce complemented by a robust operational model. This model involves hiring individuals with the requisite data and AI skills, nurturing an inherently data-driven culture, and encouraging collaborative, cross-functional teamwork.

This ensures that the organization's resources and capabilities are fully leveraged, maximizing the chances of a successful transformation. It is about creating an environment where diverse talents unite to foster innovation, efficiency, and a shared vision toward reaching new heights.

Such a model prioritizes continuous learning and adaptation. By investing in talent development and embracing a model that encourages flexibility, organizations can confidently navigate the complex terrain of digital transformation, ensuring that every team member is equipped and ready to face the challenges ahead.

Delivery Methodology

Just as climbers rely on maps, compasses, and a deep understanding of the mountain to guide their journey, organizations must adopt agile practices, iterative development, and a culture of continuous improvement. This ensures a readiness to adapt to unforeseen challenges, mirroring a mountaineer's flexibility in facing variable weather and terrain.

The essence of this journey goes beyond traditional project delivery methodologies. Organizations must foster an environment that encourages experimentation, embraces learning from failures, and champions exploratory ventures. Cross-functional teams become expedition groups by adopting a dynamic and resilient operational model akin to mountaineering teams facing the world's most challenging peaks.

They navigate the digital landscape's complexities by prioritizing rapid adaptation and mutual support over strict adherence to predetermined paths. This exploratory mindset prepares organizations to achieve immediate goals and thrive amidst the ever-evolving technological landscape, ensuring continuous growth and innovation.

Governance

In data and AI transformation, governance is the critical framework ensuring safety and ethical integrity, akin to the meticulous safety protocols in mountain climbing. Establishing robust governance practices—defining clear roles, adhering to regulations and moral standards, and addressing privacy and security—is paramount. This safeguards an organization's data assets and builds trust with stakeholders, ensuring a responsible approach to data and AI usage.

A Journey Through the Three Base Camps

Like climbers needing careful planning, teamwork, and the right gear to conquer a peak, embracing the Wisdom Spire Framework equips organizations for a successful transformation journey. It ensures that businesses are prepared for the ascent and mindful of the ethical landscape to make informed, responsible decisions that resonate with their goals and societal norms.

Base Camp 1: Operational Efficiency and Data Insight

Organizations must prioritize building a robust data infrastructure in this foundational stage and converting raw data into structured, insightful information.

- **Focus areas**:

  - **Data infrastructure and governance**: Establish a solid data foundation by implementing data governance frameworks, data management practices, and secure, scalable infrastructure. This involves setting up data integration, architecture, and storage solutions.

  - **Data quality and accessibility**: Ensure data quality by implementing data cleansing, validation, and enrichment processes. Make data easily accessible to relevant stakeholders across the organization.

  - **Process automation**: Leverage AI and automation technologies to streamline and optimize core business processes, enhancing operational efficiency.

  - **Data-driven decision-making**: Foster a culture of data-driven decision-making by democratizing access to insights and empowering employees with relevant information tailored to their roles and responsibilities.

  - **Key capabilities to develop** are data engineering, data governance, process automation, and data literacy across the organization.

- **Pitfalls to avoid**:

  - **Poor quality data**: Neglecting data quality and governance leads to unreliable insights and decision-making.

  - **Having blinders on** Siloed data and limited accessibility hinders organization-wide adoption of data-driven practices.

  - **Inefficiency**: Failure to automate manual processes results in inefficiencies and suboptimal resource utilization.

Base Camp 2: Mastery

Building upon the foundation of Base Camp 1, organizations should leverage predictive analytics to optimize business objectives and achieve mastery over their goals.

- **Focus areas**:

  - **Predictive modeling**: Develop predictive models to forecast future trends, customer behavior, market dynamics, and other relevant factors influencing business objectives.

  - **Business process optimization**: Utilize predictive insights to optimize critical business processes such as pricing, supply chain management, customer experience, product development, and more.

  - **Goal achievement**: Align predictive analytics initiatives with strategic business objectives, such as customer growth, sales optimization, market expansion, and competitive positioning.

  - **Knowledge creation**: Transition from mere insights to generating actionable knowledge to inform strategic decision-making and drive organizational learning.

  - **Key capabilities to develop**: Data science, machine learning engineering, business process optimization, and cross-functional collaboration.

- **Pitfalls to avoid**:

  - **Misalignment**: Lack of alignment between predictive analytics initiatives and core business objectives leads to disordered efforts and suboptimal outcomes.

  - **Poor quality data**: Inadequate data quality or insufficient historical data compromises the accuracy and reliability of predictive models.

  - **Not sharing insights**: Siloed implementation of predictive analytics limits the ability to derive holistic insights and optimize end-to-end business processes.

Base Camp 3: Innovation

At the pinnacle of the data and AI transformation journey, organizations strive to achieve exponential growth and industry disruption by turning knowledge into wisdom.

- **Focus areas**:

  - **Contextualized knowledge integration**: Integrate domain knowledge, business context, and data-driven insights to create a comprehensive knowledge base that fuels innovation.

  - **GenAI and knowledge creation**: Leverage advanced AI technologies to generate new knowledge, solutions, and paradigm-shifting ideas.

- o **Industry disruption and thought leadership**: Leverage the organization's knowledge assets to drive industry-wide innovation, establish thought leadership, and shape the market's future.

- o **Growth and scale**: Capitalize on the transformative power of AI and data to achieve exponential growth, scale operations, and expand into new markets or venture into adjacent industries.

- o **Key capabilities to develop**: Advanced AI and ML, domain expertise, innovation management, and a culture of continuous learning and exploration.

- **Pitfalls to avoid**:

  - o **Tech just for tech's sake**: Failure to integrate domain knowledge and business context leads to disconnected or impractical AI-driven solutions.

  - o **Lack of resources**: Not investing in the infrastructure or talent for advanced AI technologies and knowledge-creation initiatives.

  - o **Fear of change**: Resistance to disruptive innovation or a need for agility adapting to rapidly evolving market conditions.

By mastering each base camp sequentially, organizations can ensure a solid foundation, optimize their processes, and ultimately achieve transformative innovation and exponential growth through the strategic application of data and AI technologies.

The journey to AI maturity is intricate and nuanced, with only a tiny fraction of companies reaching the pinnacle of their potential. Despite the availability of technologies like GenAI and predictive modeling, it is surprising that less than 10 percent of companies achieve what can be called "peak performance"—the stage where they realize their true potential and experience exponential growth. This disparity highlights the delicate balance between technology and strategic application and the importance of prioritizing foundational capabilities at each stage of the AI adoption journey.

The critical point is not when to apply specific AI technologies but the need to master foundational capabilities at each base camp before attempting to ascend to the next level. This approach ensures a solid foundation, optimizes processes and implements transformative innovation and exponential growth.

While GenAI or predictive modeling can be applied from the beginning, neglecting to prioritize and develop core competencies like data infrastructure, governance, and literacy at Base Camp 1 can hinder an organization's ability to progress to Base Camp 2, where the focus shifts to leveraging predictive analytics for business optimization. Similarly, skipping the critical step of aligning predictive initiatives with core objectives at Base Camp 2 can impede the successful transition to Base Camp 3, where the goal is to achieve industry disruption and exponential growth through knowledge creation and AI-driven innovation.

The journey is not about the specific technologies employed but the strategic and sequential development of capabilities that enable organizations to harness the full potential of AI and data. This systematic approach separates the minority of successful AI leaders from the majority of AI explorers who often struggle to move beyond the initial experimentation phase or scale their efforts for transformative impact.

Navigating Treacherous Slopes

Organizations encounter obstacles just as mountain climbers confront unpredictable weather and treacherous terrains. These challenges, if addressed, can ensure progress and lead to a transformative endeavor. **Here are some of these challenges and strategies for overcoming them:**

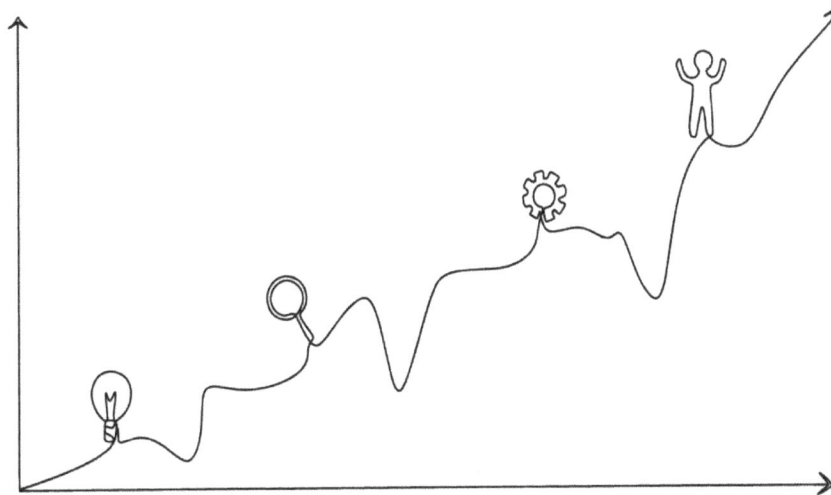

Figure 23. Navigating the River of Transformation

Uncharted Territories

Organizations often need help with unfamiliar data sources, technologies, and methodologies, akin to climbers venturing into uncharted terrain. This unfamiliarity can lead to missteps, such as adopting technologies that do not align with business goals.

- **Example**: Blockbuster's failure in the 2000s to adapt to the digital streaming trend while Netflix was doing so (B4AD56 2018; O'Brien 2023).

- **Critical insight**: Strategies for continuous learning and staying updated with the latest in data science and AI are essential. Specific approaches, like regular training sessions and workshops, can help familiarize teams with new terrain.

The Altitude Challenge

Organizations face budget, talent, and time constraints. There are resource management challenges akin to climbers rationing their supplies. A relevant example is the healthcare sector, where AI is expected to revolutionize care delivery and workforce dynamics. This innovation comes as the demand for healthcare services grows while a shortfall of professionals looms. Efficiently leveraging AI can address these gaps, enhancing care and operational efficiency.

- **Example**: In 2017, the healthcare provider Sutter Health faced challenges leveraging AI for predictive diagnosis due to a lack of in-house expertise. They collaborated with external AI specialists to develop models that could predict clinical events, significantly improving diagnostic accuracy and patient outcomes (Spatharou, Hieronimus, & Jenkins, 2020; Choi, Bahadori, & Schuetz, 2015).

- **Critical insight**: It is crucial to prioritize projects based on ROI and efficiently allocate resources. Phased implementation can help manage constraints effectively.

Unpredictable Weather

Internal resistance within organizations can slow down or derail transformation efforts, much like climbers face environmental resistance on their ascent.

- **Example**: Nokia's inability in the late 2000s to adapt quickly to the smartphone revolution (Ishalli 2023).

- **Critical insight**: Effective change management strategies emphasizing communication and leadership are paramount. Practical insights from companies that successfully manage internal resistance can provide valuable lessons.

Lack of a Clear Vision

Without a clear vision or strategy for their data/AI transformation, organizations may invest in initiatives that fail to drive meaningful results. This situation resembles climbers on an expedition without a defined route or summit goal, leading to wasted resources and missed opportunities to reach new heights.

- **Example**: Yahoo's multiple business strategy shifts in the 2000s without a clear vision (Almalki 2020).

- **Critical insight**: Establishing a clear data/AI strategy aligned with broader objectives is essential. Setting measurable goals ensures alignment and progress tracking.

Technical Challenges

Navigating technical challenges is akin to rocky patches and crevasses in mountainous terrain. Organizations often encounter obstacles such as data silos and integration issues.

- **Example**: HSBC's difficulties in the 2010s were influenced by issues related to legacy systems, modernization, and the strategic integration of AI technologies (Flinders 2014).

- **Critical insight**: Highlighting the criticality of robust IT infrastructure and effective strategies for integrating new technologies with legacy systems, such as leveraging microservices architecture or cloud solutions, is paramount for successful implementation.

Ethical Crevasse

Organizations must navigate the ethical implications of AI, from algorithm bias to data privacy concerns. This challenge is similar to facing a crevasse while climbing a mountain, where a misstep can lead to significant setbacks or even failure.

- **Example**: In 2018, Amazon's AI recruiting tool showed bias against female candidates (Vincent 2018).

- **Critical insight**: Highlighting the importance of ethical considerations in AI development and deployment, alongside compliance with regulations. Transparency and addressing biases in algorithms are crucial.

Organizations can successfully navigate the treacherous slopes of data/AI transformation by addressing these challenges with targeted strategies.

Learning from the Trailblazers

As organizations embark on their transformational journies, there is much to learn from those who have trodden the path before. Here are real-world anecdotes of companies that faced challenges head-on, learned from their setbacks, and emerged stronger:

- **Netflix**

  - **Positive anecdote**: In 2017, Netflix introduced interactive storytelling (e.g., *Black Mirror: Bandersnatch*), allowing viewers to make choices that affect the story's outcome, thereby showcasing innovative content engagement (Engelbrecht 2017).

  - **Negative anecdote**: Netflix faced backlash 2011 when it announced the separation of its DVD rental and streaming services, leading to a short-lived rebranding of its Qwikster (Wolverton 2011; Chappell 2011).

- **Target**

  - **Positive anecdote**: Target used predictive analytics 2012 to tailor marketing strategies (Hoag 2012).

  - **Negative anecdote**: In 2013, Target suffered a significant data breach, compromising the personal information of millions of customers, which underscored the critical need for robust data security measures (Krebs on Security 2015; Young 2021).

- **Apple**

  - **Positive anecdote**: Apple launched the App Tracking Transparency feature in 2021, enhancing privacy by allowing users to control which apps can track their activity across other companies' apps and websites (Collective Measures 2021).

  - **Negative anecdote**: Apple Maps' initial release in 2012 was met with widespread criticism due to its inaccuracies and lack of detailed information, prompting a public apology from then-CEO Tim Cook (Bonnington 2012).

- **Amazon**

  - **Positive anecdote**: In 2018, Amazon's Alexa became one of the leading examples of successful consumer AI with its vast skills and capabilities, from playing music to controlling smart home devices (Marr 2018).

  - **Negative anecdote**: Amazon faced controversy over AI recruiting tool biases in 2018, with reports indicating significant gender bias in the algorithm affecting women applicants. The company took steps to address these, but concerns remained about the discriminatory nature of the tool (Zheng, 2023).

- **Microsoft**

  - **Positive anecdote**: Microsoft's Azure AI achieved a significant milestone in 2017 by generating more accurate speech recognition, surpassing human levels of understanding in specific benchmarks (Nemire 2016).

  - **Negative anecdote**: Microsoft's AI chatbot Tay was quickly exploited in 2016 by Internet users, who trained it to generate offensive and discriminatory responses. This highlighted the importance of careful monitoring and the potential for AI to be misused (Kraft 2016).

These anecdotes showcase the highs and lows experienced by trailblazing companies, emphasizing that even industry leaders can make mistakes. While the Wisdom Spire Framework provides a structured approach, additional frameworks, and proactive measures can further guide organizations in their ascent.

Conclusion

The critical role of conscious engineering and a deep understanding of models of reality led to the implementation of the Wisdom Spire Framework to guide organizations through data and AI transformation journeys. By challenging existing beliefs and embracing these principles, data, and AI can be seen as tools and catalysts for meaningful innovation aligned with values. This journey is more than technological—it is a comprehensive process that reimagines innovation, leadership, and growth.

PART TWO

Transformation In Action

CHAPTER 7

The Summit

Imagine standing at the base of a towering mountain. Not just any hill but a representation of an organization's future in the digital age. The peak is shrouded in clouds, representing the uncertainties and infinite possibilities of data and AI transformation. Climbing will be powered by a clear vision (the "why") and the team's collective strengths and aspirations (the "who"). The journey is about reaching the summit by discovering new paths and seeing new vistas—the essence of transforming with AI and data.

Standing ready to ascend this digital peak, it is time to consider the climb as part of a holistic growth and renewal journey. Reflecting the natural stages of human development, **organizational transformation can be segmented into five key phases:**

1. **Decluttering**: Assessing the current landscape and shedding outdated practices to align with the core mission better.

2. **Visioning**: Crafting a clear and strategic vision regarding how data and AI support and drive business objectives.

3. **Communication**: Ensuring that the vision is understood and embraced across all levels of the organization to foster alignment and commitment.

4. **Action**: Implementing prioritized initiatives to demonstrate commitment to this new strategy.

5. **Receiving**: Cultivating a culture of openness to learn from success and fuel the ongoing journey by overcoming challenges.

This structured approach to transformation is the preparation necessary to tackle the climb and embrace the myriad paths and discoveries that lie ahead.

A well-defined strategy is the map and compass to guide an organization toward its goals. It provides a framework for decision-making, ensuring that every decision—simple or critical—aligns with the overarching strategic objectives. This straightforward strategy will allow for meaningful discussions, well-informed choices, and the effective prioritization of initiatives—opening up new possibilities for innovation and growth.

For example, consider Netflix's recommendation engine. It is a prime instance of the transformative potential of a data-driven strategy. The system, far from being a mere feature, was a strategic asset deeply embedded into Netflix's overarching approach to enhancing user engagement and satisfaction. By analyzing extensive data on user preferences, viewing times, and content popularity, Netflix leveraged sophisticated algorithms to personalize content recommendations for each user.

This data-driven approach revolutionized the user experience significantly in terms of viewer retention and subscription growth. It demonstrates how a focused data and AI strategy can turn user data into a competitive advantage (Netflix Technology Blog, 2013).

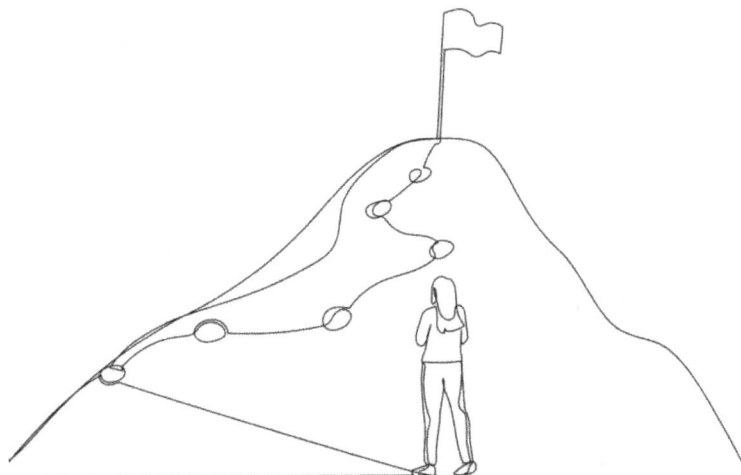

Figure 24. Transformation as a Mountain Journey

As I wrote in the last chapter, Simon Sinek's *Start With Why* offers a thought-provoking perspective on the importance of purpose and provides practical guidance on how to drive personal and organizational success. Sinek introduces the concept of the Golden Circle, which consists of three layers: the why, the how, and the what. I want to build on Sinek's concept and explain how organizations and businesses can develop data and AI strategies by addressing why, what, and how.

The interwoven threads of why and who form the foundation of data and AI strategy. Beyond the mechanics of analysis and plotting paths, a team's vision and collective drive will anchor the journey.

While best practices provide a solid foundation for technological advancement, my approach integrates the profound dimensions of why (purpose) and who (people) in data and AI initiatives. This chapter is not just about navigating the technicalities of data management and AI application; it is about infusing these endeavors with a sense of mission and collaborative spirit. This perspective melds strategic rigor with transformative vision, a departure from mainstream data and AI strategy narratives.

Defining the Heart of Transformation

An organization should articulate a clear and compelling vision for how data and AI can transform operations to ensure alignment with the overall business strategy. This vision should highlight data and AI's value to organizational objectives, define the desired future state, and outline the impact to be achieved through planned initiatives. This guiding vision inspires and motivates employees, instilling a sense of purpose and fostering a collective drive to work towards a common goal.

Aligning Vision with Action

Transformative success is not just paved with raw data and AI but by harmonizing the why with the who. Like a well-danced flamenco, aligning a data and AI strategy with broader business aspirations requires precision and passion. It invites us to consider how data and AI serve operational goals and resonate with the mission to engage customers, innovate productively, and enhance operational excellence.

Such alignment transcends the mere technical integration of new technologies into workflows. It needs to be a symbiotic ecosystem where every stakeholder, led by visionary leaders, becomes a champion for transformation. The why and who serve as twin beacons to guide aligning strategy with core business objectives, embedding data and AI not as external tools but as intrinsic elements of an organization's DNA.

For instance, if the data strategy does not explicitly support a business's objective to enhance customer satisfaction, this misalignment is a red flag that should prompt an immediate re-evaluation to ensure that the strategic focus is firmly on driving customer-centric outcomes.

Cultivating Transformation Through Pragmatism

Pragmatic steps will bring such transformation to life. It is grounded in the core why (the purpose that propels us forward) and empowered by the who (the team's collective talents and aspirations), data, and AI transformation becomes a journey of meaningful progress and tangible outcomes. It is not defined by rigid plans but by adaptive, actionable strategies that resonate with team dynamism and organizational ethos.

Data/AI Vision Statement

Specific, measurable goals and objectives must be defined to provide clear direction. Strategic objectives can focus on enhancing customer experiences, reducing costs through automation, mitigating risks, or driving new revenue streams. Prioritize objectives based on feasibility, business impact, and resource requirements.

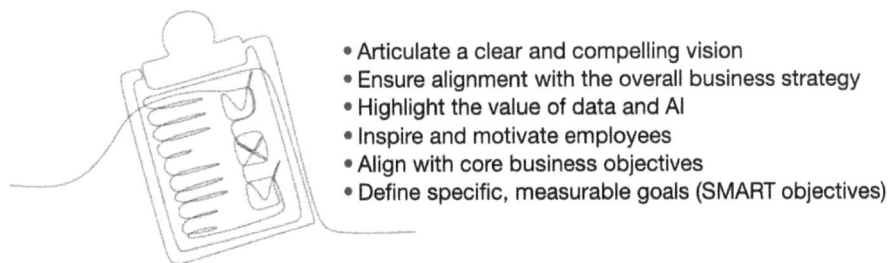

- Articulate a clear and compelling vision
- Ensure alignment with the overall business strategy
- Highlight the value of data and AI
- Inspire and motivate employees
- Align with core business objectives
- Define specific, measurable goals (SMART objectives)

Figure 25. Checklist for Defining TO-BE

Assessing the Current Landscape (What)

Instead of static assessments, I embrace a philosophy of continuous exploration and growth, an evaluation of data and AI capabilities that becomes an ongoing dialogue within the organization—one that is less about reaching a fixed maturity level and more about constantly adapting and evolving. This adaptive assessment approach encourages viewing maturity not as a destination but as a journey marked by continuous learning, iterative improvement, and alignment with the mission.

Shifting from conventional maturity models to a more fluid framework recognizes each organization's unique path. It is a framework that supports commitment to transformative change and urges asking, "How can our current capabilities serve our purpose more effectively?" and "How can we empower our team to drive this evolution?" Treating data and AI maturity as an organization's living, evolving aspect can be aligned more closely with core values and the dynamic digital landscape.

- **Decluttering as a strategic tool for clarity**: In the spirit of continuous adaptation, I draw upon Marie Kondo's decluttering ethos and extend its principles from personal tidying to organizational focus (KonMari, n.d.). This approach involves critically examining and simplifying existing data and AI processes to ensure they directly contribute to strategic goals and truly "spark joy"—in this context, a tangible value and alignment with core business KPIs.

- **Evaluating current practices**: Begin by assessing each data, AI process, and system currently in use. Question their relevance and utility in the current business context, like deciding which items to keep in a tidy home. This evaluation should focus on whether these processes support or hinder strategic objectives.

- **Identifying attachments and overcoming fear**: Address common hurdles such as attachment to familiar but outdated systems and apprehension towards new, untested technological solutions. Strategically, this involves guiding the organization in letting go of obsolete practices and embracing innovative approaches that promise more significant benefits.

- **Aligning with joy**: Redefine what brings value and happiness within the organization by connecting every process and decision to the overarching business KPIs and growth. This step ensures that only the most impactful and efficient practices remain in place, streamlining operations and focusing on driving success.

Case Study: Company X's Data Strategy Decluttering

In one of my past projects, I worked with Company X, an anonymized multinational manufacturing company, to carry out a data strategy decluttering initiative. The company had been operating with siloed systems and inefficient reporting processes for years. By evaluating each process through core KPIs like inventory efficiency and supply chain optimization, we could discontinue redundant activities and implement a unified data platform. This decluttering initiative enabled significant cost savings, improved decision-making speed, and created better alignment with their strategic goals.

The Decluttering Process

- **Comprehensive assessment**: We began by assessing each data, AI process, and system currently in use, questioning their relevance and utility in the current business context. This step was akin to deciding which items to keep in a tidy home. The evaluation focused on whether these processes supported or hindered strategic objectives.

- **Breaking down data silos**: Our strategy involved integrating siloed systems into a unified data platform. This integration was crucial for improving data flow and ensuring that all data could be analyzed holistically rather than in isolated segments. This process helped eliminate redundancies and streamline operations.

- **Utilizing advanced analytics**: We employed advanced analytics and machine learning to evaluate process performance and efficiency. By analyzing data from various sources, we could identify areas where processes were not contributing to overall efficiency and productivity. This information informed our decisions on which processes to retain and eliminate.

- **Continuous improvement philosophy**: Decluttering was not a one-time activity but a continuous process. We adopted a business philosophy of ongoing evaluation and adaptation. This approach ensured that the company's data and AI strategy aligned with evolving business goals and the dynamic digital landscape.

- **Customer-centric approach**: Our strategy also included a strong focus on customer collaboration. By aligning processes with customer needs and expectations, we ensured that data and AI initiatives provided tangible value and supported long-term strategic objectives.

By breaking down data silos and creating a more integrated data platform, Company X achieved significant cost savings, improved decision-making speed, and better aligned its operations with strategic goals. The continuous decluttering of data systems is now a core aspect of Company X's business philosophy. It ensures ongoing alignment with an adaptive assessment model and keeps the organization agile and focused on its mission.

Activity:

Initiate a decluttering audit of existing data and AI practices. Rigorously evaluate each process, system, and technology to determine whether it aligns with and directly contributes to core strategic objectives. Let go of aspects that no longer "spark joy" in creating tangible value in streamlining the technological landscape.

Gap Analysis: Where You Are Versus Where You Want to Be

Conduct a comprehensive assessment of the organization's current data capabilities, AI readiness, and existing data infrastructure to identify strengths, weaknesses, and gaps and provide a baseline for future improvement. Based on the vision and current state assessment, define strategic objectives that outline what the organization aims to achieve through data and AI initiatives. These objectives should be specific, measurable, attainable, relevant, and time-bound (SMART) and align with the overall business goals.

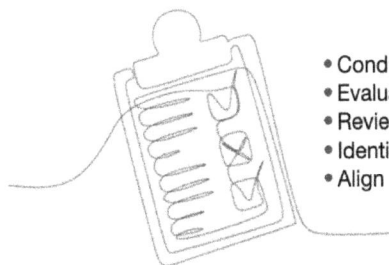

- Conduct a comprehensive assessment of data capabilities
- Evaluate AI readiness
- Review existing data infrastructure
- Identify strengths, weaknesses, and gaps
- Align with the overall business goals

Figure 26. Checklist for Defining AS-IS

Mapping the Journey

When developing a roadmap for data and AI transformation, treat it not just as a document but as a living, breathing strategy deeply influenced by your why (purpose) and who (people). Such an approach transcends traditional planning, ensuring the path forward is dynamic and reflects an organization's evolving needs and ambitions.

Each initiative within the roadmap is deliberately aligned with the organizational why, embedding the mission to innovate and transform into every strategy and action. This ensures that the journey through the complexities of data and AI transformation is not just about technological advancement but about fulfilling a broader purpose that resonates with every team member and your organization's core values.

Central to this adaptive roadmap is the who—the dedicated teams and individuals whose passion and expertise bring the overall vision to life. Initiatives are selected for their technical merits and ability to empower people. These will foster an environment where everyone can contribute and benefit from the transformative journey. Investments in technology, skills development, and agile methodologies are made to support the team's growth and adapt to their feedback and insights.

Navigating with Flexibility and Insight

The dynamic nature of the digital landscape necessitates an inherently flexible and responsive roadmap. Strategies should accommodate regular reassessment and realignment based on new insights, technological advancements, and the organization's evolving needs. This adaptable approach ensures that strategic initiatives remain relevant and impactful and effectively guides efforts toward long-term objectives.

Integrating continuous learning and realignment mechanisms will be a compass that continually orientates purpose. This evolving roadmap is as responsive as it is deliberate, ensuring that data and AI initiatives are continuously shaped by and for people and move towards the mission's overarching purpose.

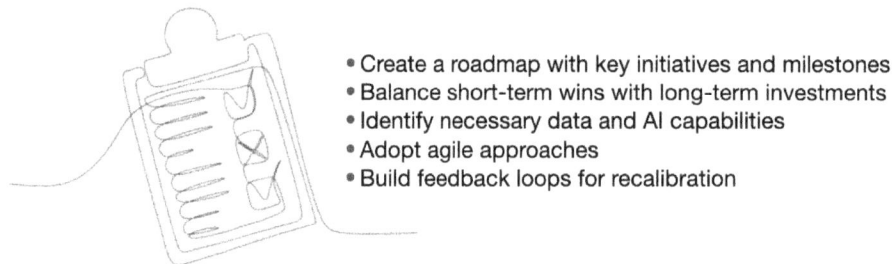

- Create a roadmap with key initiatives and milestones
- Balance short-term wins with long-term investments
- Identify necessary data and AI capabilities
- Adopt agile approaches
- Build feedback loops for recalibration

Figure 27. Checklist for Creating a Roadmap

Offense and Defense: Strategic Equilibrium

In navigating the transformative landscape of data and AI, drawing inspiration from strategic business planning and the principles of spiritual growth offers a unique perspective on achieving success. Balancing offensive and defensive strategies, as highlighted in the seminal *Harvard Business Review* article "What is Your Data Strategy?" by Thomas H. Davenport, provides a foundational framework for this discussion (Davenport 2017). This balance is crucial, not just for the operational efficacy of data management and usage, but also reflects an organization's core purpose (why) and the collective strength of its team (who).

The Defensive Foundation

A defensive strategy is the bedrock to build secure and compliant data practices. Operational efficiency and data integrity testify to the purpose driving the organizational mission. This strategy emphasizes safeguarding data assets, ensuring regulatory compliance, and harnessing data for operational efficiency and risk mitigation, all of which directly contribute to organizational objectives and values.

- **They are building a solid foundation (why)**: The defensive strategy is rooted in maintaining the highest data privacy, security, and quality standards—reflecting an organization's core values of trust and responsibility toward stakeholders. Specific practices such as anonymization, encryption, and real-time security monitoring are ways to protect data assets. Adhering to regulations like the EU's General Data Protection Regulation (GDPR) or the California Consumer Privacy Act (CCPA) demonstrates compliance while enhancing data security and customer trust.

- **Empowering quick team wins (who)**: A team's expertise and dedication power the implementation of the defensive strategy. The team plays a crucial role in demonstrating the tangible benefits of data-driven initiatives by focusing on areas where quick wins can be achieved—such as process optimization and automation. This builds an organization's confidence and trust and fosters continuous improvement and innovation, swiftly responding to changing regulations and market dynamics. For example, a healthcare organization's adoption of data encryption and access controls to protect patient data and align with the US Health Insurance Portability and Accountability Act (HIPAA) showcases how a defensive strategy can enhance trust and operational efficiency.

Incorporating why and who into a defensive strategy underlines the importance of building a solid foundation based on organizational values and a team's collective capabilities. It showcases how defensive data and AI practices are not merely about compliance and risk management but also integral to realizing the mission and empowering a team to drive immediate and impactful changes.

Offensive Capabilities

The offensive segment is an organization's vision for the future (why) and collective strength (who) coalescing to pioneer innovation, drive growth, and redefine the industry landscape.

- **Pioneering through vision (why)**: The offensive strategy manifests the why—the core mission to participate in the market and lead and redefine it through innovation. This strategy involves leveraging data and AI to identify new opportunities to create unique value propositions and deliver unparalleled customer experiences. By aligning this strategy with an organization's vision, every innovative endeavor is a step towards realizing a future where the business thrives and sets new standards of excellence and engagement. Creative uses of AI and big data, such as predictive analytics for customer behavior and AI-driven personalization strategies in marketing, exemplify this.

- **Collaborative innovation (who)**: The success of an offensive strategy is propelled by the who—the organization's diverse, talented individuals. Their creativity, expertise, and collaborative spirit translate visionary ideas into tangible solutions. By fostering a collaborative environment, a team is empowered to take calculated risks, learn from outcomes, and continuously push the boundaries of what is possible. Collective effort across teams, from data scientists to marketing and product development, is crucial in realizing this vision.

- **Strategic impact**: The intersection of why and who ensures the pursuit of innovation is purposeful and people-driven, leading to breakthroughs that can transform an industry and strengthen an organization's internal capabilities and culture.

When navigating the challenges and opportunities of data and AI transformation, the offensive strategy serves as a beacon that guides efforts toward impactful innovations. The tangible benefits of such a strategy include market share growth, entry into new markets, or the development of new revenue streams—all aligning with the company's why and achieved through the collective genius of the who.

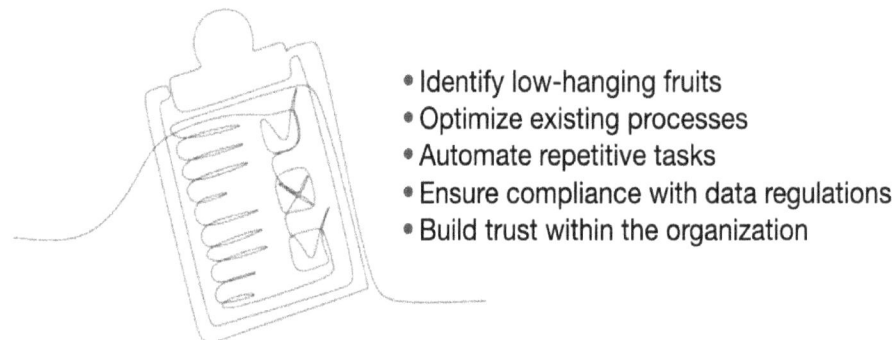

- Identify low-hanging fruits
- Optimize existing processes
- Automate repetitive tasks
- Ensure compliance with data regulations
- Build trust within the organization

Figure 28. Checklist for Defensive Data and AI Strategy

Integrating Strategies with Purpose

Challenges and opportunities lie in harmonizing these strategies and ensuring that data and AI initiatives are secure, compliant, and innovative. By thoughtfully integrating defensive and offensive strategies, the journey becomes a balanced pursuit of excellence, innovation, and ethical responsibility.

Pioneering Through Purpose

Venturing into proactive innovation and growth means harnessing an offensive data and AI strategy as a plan and manifestation of an organization's courage to pioneer. It is a strategy that does not just react to trends but seeks to create them, driven by the vision to bring about meaningful change and growth.

- **The role of why**: The offensive strategy is a bold declaration of a commitment to adapting and shaping the future. It reflects a deep understanding of the transformative power of data and AI, harnessed not merely for incremental improvements but for groundbreaking innovations that align with the organizational mission. This vision-driven approach ensures that every initiative under the offensive strategy is a step toward realizing broader goals.

- **The strength of who**: Behind every technological advancement and strategic pivot are the people who make it happen. An offensive strategy thrives on the team's creativity, expertise, and passion. This people-centric focus ensures that growth and leadership are propelled not just by data and algorithms but by the inspired actions of people.

Embracing such a forward-thinking strategy requires more than just technological investment; it demands nurturing an environment where your why and who can thrive together. In this synergistic space, an organization can truly distinguish itself by driving innovation that's not only technologically advanced but deeply resonant with purpose. Data and AI are transformed from mere tactics to strategic movements that chart new territories of opportunity and success.

Nurturing Your Why through New Models

The offense strategy embodies more than just the pursuit of competitive edge—it manifests the why, the core mission to innovate and grow responsibly. This strategy identifies opportunities where data and AI can create immediate value and propel an organization toward sustainable, mission-aligned growth.

- **Personalized experiences (the why)**: Consider implementing a recommendation engine on an e-commerce platform. This is not just a technical endeavor but a strategic move to deepen customer engagement and satisfaction, reflecting a commitment to providing value that's both personalized and impactful. It is not just about increasing sales. A shopping experience that resonates with customers' needs and preferences is nurtured and directly reflects the way in action.

- **The power of conversational AI (the who)**: Deploying a chatbot for customer support transcends technological innovation; it is about enhancing the quality of customer interactions. This initiative reflects the team's () dedication to improving service efficiency and accessibility, ensuring every customer receives timely and practical support. Through initiatives like these, data and AI strategy becomes a living expression of the mission, with the team playing a crucial role in bringing this vision to life.

When exploring these new business models and technologies, remember that each step forward is an opportunity to reinforce your why (the purpose) and to celebrate the who (the team behind these innovations). By aligning offensive strategies with the organization's mission and leveraging the collective team strengths, the journey is not just about leading the market but about creating a meaningful, lasting impact that mirrors the values and aspirations of the organization.

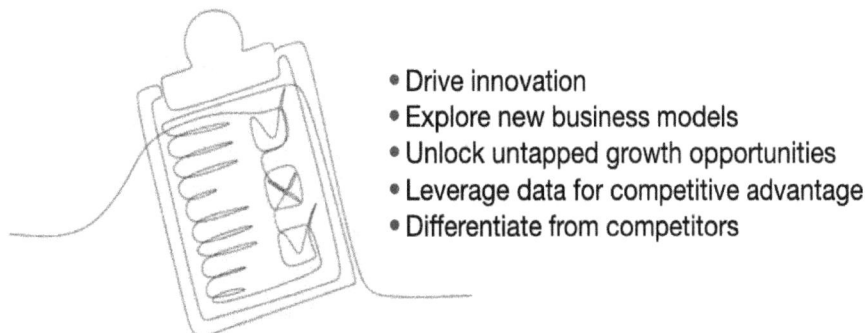

- Drive innovation
- Explore new business models
- Unlock untapped growth opportunities
- Leverage data for competitive advantage
- Differentiate from competitors

Figure 29. Checklist for Offensive Data and AI Strategy

Finding Balance

The key challenge in data and AI strategy is finding the right balance between offense and defense. While quick wins and optimization provide immediate business impact, more than focusing on them can lead to scattered innovation efforts that fail to achieve true transformation. Conversely, only pursuing long-term investments in technology and platforms without creating immediate value can erode trust and hinder organizational adoption. The key is to strike a balance that delivers concrete business use cases while laying the foundation for long-term transformation.

Just like climbers taking small steps while ascending the mountain, organizational transformation must be broken down into manageable increments. These small steps represent the tasks, initiatives, and use cases that collectively contribute to reaching the summit. Through such small steps, the balance between offense and defense is found.

The defense strategy is like building a solid foundation for a mountain ascent. Necessary precautions are taken, such as setting up base camps, acquiring the right equipment, and ensuring the safety of climbers. In a data and AI transformation, the defense strategy involves optimizing existing processes, automating repetitive tasks, and ensuring compliance with data regulations. Just as climbers secure their safety ropes and take precautions before advancing, prioritizing quick wins and optimization builds trust and credibility within an organization.

However, an offensive strategy is also needed to push the boundaries of innovation and growth, and it represents the drive to reach new heights and conquer uncharted peaks. This involves taking calculated risks, exploring alternative routes, and embracing challenges.

Case Studies

- **Examples of offensive and defensive strategies**:
  - **Amazon's recommendation engine (success)**: Powered by years of accumulated user data, Amazon's recommendation engine has been a cornerstone of its growth strategy. By suggesting relevant products to customers based on purchase history and browsing behavior, this offensive data strategy contributes significantly to sales, with recommendations influencing about 35 percent of purchases. The recommendation system's success is due to sophisticated algorithms that include item-to-item collaborative filtering, which effectively maps customer preferences to product features, enhancing the shopping experience and driving incremental sales (Gokul, n.d.).
  - **Decline of Friendster (failure)**: Friendster's decline is a cautionary tale of a defensive strategy gone awry. Once a popular social media platform, Friendster failed to innovate and leverage data to enhance the user experience and engagement. Their focus on maintaining existing features rather than driving growth through data insights contributed to their eventual decline, as they lost ground to competitors who effectively used data to adapt and evolve (Velazco n.d.

- **Examples of quick wins versus long-term innovation**:

  o **Uber's surge pricing (success)**: Introduced in the mid-2010s, this is an example of a quick win through offensive strategy. By leveraging real-time data on rider demand and driver supply, Uber implemented adaptive surge pricing to balance supply-demand dynamics, significantly improving service reliability and contributing to its growth as it expanded its core service offerings (Dholakia 2015).

  o **Yahoo's innovation struggle (failure)**: Yahoo's failures highlight the pitfalls of not pursuing a robust offensive data strategy when it relied on partnerships to develop innovative products and AI capabilities, particularly in the late 2000s and early 2010s. For example, Yahoo entered a 30-year strategic alliance with Taboola to power Yahoo's recommendations (Taboola 2023). Yahoo's missed opportunities, mismanagement of acquisitions, and lack of clear vision and leadership led to its decline compared to Google, which focused on innovation, collaboration, and user-centric strategies (Robert H. Smith School n.d.).

- **Examples of data-driven decision-making**:

  o **Netflix's *House of Cards* (success)**: The decision to produce *House of Cards* is a testament to successful data-driven decision-making. By analyzing viewer data, Netflix identified a strong demand for a remake of the British political drama. Premiering in 2013, it became one of their most successful original series. This strategic use of data to inform content creation has been a critical factor in Netflix's rise as a dominant force in the streaming industry (Sachdeva, 2023; Sadeh, 2019).

  o **Blockbuster's downfall (failure)**: by the late 2000s, Blockbuster's demise can be partly attributed to not leveraging data on evolving consumer preferences. This starkly contrasts with Netflix's data-driven approach, which helped them stay ahead of industry trends and customer needs (Satell 2014).

- **Examples of AI-driven decision-making**:

  o **Mayo Clinic's AI integration (success)**: Assisting radiologists, particularly from the 2010s onwards, with AI-driven algorithms that primarily focus on clinical decision-making, cardiovascular medicine, disease detection, and early disease detection using patient data and AI demonstrates the positive impact of forward thinking. Mayo Clinic's AI improved the speed and accuracy of diagnoses and patient outcomes (Lee & Yoon 2021).

  o **The challenge of Watson for Oncology (failure)**: Introduced to aid cancer treatment, IBM's Watson for Oncology showed promise by supporting clinicians with evidence-based treatment options. While it showed promise, its effectiveness was mixed, with some recommendations deviating from best practices. These were identified and corrected during IBM's stringent testing, underlining the importance of continual refinement and validation in AI applications in healthcare (Luxton 2019).

- o **Amazon's inventory optimization (success)**: Utilizing AI algorithms to optimize inventory management by analyzing purchasing behaviors, seasonal trends, and warehouse locations, Amazon's system forecasts which items will likely be bought in specific geographic locations and at certain times. This allows for optimized inventory levels and reduced costs, furthering efficiency in operations (Supply Chain Today, n.d.).

- o **Air Canada's chatbot controversy (failure)**: A customer service chatbot intended to automate responses to common queries faltered when the bot failed to handle complex or sensitive issues appropriately. The bot's insufficient training in these areas led to inappropriate responses to customer complaints, sparking a legal dispute and undermining customer trust. This failure underlines the importance of tailoring AI applications to handle the nuances of customer interactions (Loe, 2024; Cecco, 2024).

The offense strategy encourages the exploration of new business models, unlocks untapped opportunities, and drives sustainable growth. Just as climbers push their limits to scale new peaks, leveraging data and AI aims to disrupt the market and differentiate companies from competitors.

The harmonious integration of offense and defense is paramount. The balance should not merely be a statement of intent but a practiced and nuanced art, something I have come to understand through my professional experiences.

In working with organizations, I have seen the pitfalls of a defense-only approach. Companies become mired in maintaining the status quo, which leads to stasis and obsolescence. Conversely, a relentless pursuit of offense without regard for foundational stability can result in misaligned initiatives and squandered resources. Imbalances can erode stakeholder trust and stifle innovation.

Reflecting on these experiences, I would like to advocate for a strategy that marries the quick wins of defense with the ambitious vision of offense. It is a strategy that prizes the steady accumulation of value from data and AI as much as it does the disruptive potential of new technologies. For instance, at one organization I worked with, the overemphasis on aggressive market capture through AI led to a neglect of data quality, ultimately compromising the system's reliability. In contrast, another client's overly cautious approach to data governance stifled their ability to leverage AI for market insights, leaving them behind the competition.

By judiciously alternating focus between these strategies, organizations I have advised have been able to course-correct towards more balanced, practical approaches. They have learned that the strategic interplay of defense and offense is not just a theory but a lived practice that demands ongoing reassessment and recalibration.

A successful data and AI strategy is dynamic. It evolves with the organization's needs and the ever-changing digital landscape. The ultimate goal remains a steadfast progression toward a robust, data-enlightened future—with the summit continually redefined by the journey itself.

Embracing Surrender

Like an artist finding harmony between chaos and order, balancing the innovation equilibrium enables organizations to thrive in the digital age. The essence of surrender—letting go and trusting the process—becomes a guiding principle for personal growth and organizational innovation, allowing businesses to balance offensive and defensive data/AI strategies.

Figure 30. Surrender

The Essence of Surrender: Mastery Over Chaos

Surrender in the context of data and AI involves a strategic pivot from merely amassing data to transforming it into actionable insights. This transformation is achieved by journeying through the layers of data, turning it into information, cultivating knowledge, and ultimately distilling wisdom. It is about building an internal scaffolding of understanding to empower the navigation of AI complexities with finesse and insight.

- **Data to information:** The first step is organizing raw data's chaotic abundance into coherent, structured information. This involves cleaning, categorizing, and contextualizing data to make it accessible and understandable.

- **Information to Knowledge:** With structured data, the journey continues toward synthesizing knowledge. Analytical techniques and ML algorithms extract patterns, trends, and correlations to connect the dots and unveil hidden narratives.

- **Knowledge to wisdom:** The pinnacle is the leap from knowledge to wisdom. Wisdom is applying insights ethically, sustainably, and in alignment with humanity's broader goals to reflect the profound implications of AI technologies.

Finding Equilibrium through Surrender

The offensive strategy focuses on driving innovation, exploring new business models, and achieving long-term growth. It requires surrendering to the unknown, embracing uncertainty, and being open to disruptive possibilities that emerge from data-driven insights. Conversely, the defensive strategy centers around creating immediate value, optimizing existing processes, ensuring compliance, and harvesting the low-hanging fruit. Here, surrender manifests as the willingness to let go of outdated practices and adapt to data-driven efficiency and risk-mitigation recommendations.

Figure 31. The Strategic Equilibrium between Offense and Defense

Striking the right balance between these approaches is crucial for sustainable transformation. By embracing the principle of surrender, organizations can navigate this delicate equilibrium with agility and wisdom, leveraging data and AI to unlock new growth frontiers while maintaining a solid foundation for stability and compliance.

The Transformative Power of Exploratory Data Analysis

One of the oft-underrated promises of data science lies in the transformative potential of exploratory data analysis (EDA), which embodies the spirit of surrender by encouraging analysts to approach data openly, free from preconceived notions or rigid hypotheses. By surrendering to the insights delivered, organizations can uncover hidden patterns, unexpected correlations, and novel opportunities that traditional, hypothesis-driven approaches may overlook.

The EDA journey mirrors personal growth, where individuals must let go of limiting beliefs and embrace the unknown to unlock their full potential. Organizations can harness the power of EDA to navigate the complexities of data and AI and drive innovation.

Adapting and Innovating through Surrender

The evolution of AI serves as a powerful metaphor for the transformative potential of surrender. From the early enthusiasm for symbolic AI to the challenges of the AI winters, the field progressed by surrendering outdated methodologies and embracing new paradigms like machine learning and deep neural networks.

114

Similarly, in data science, the transition from rigid statistical models to flexible, data-driven approaches like machine learning reflects the importance of letting go of preconceived notions and being open to new possibilities. This willingness to learn from unbiased data, challenge established norms, and embrace empirical evidence over doctrine mirrors the essence of surrender in personal growth.

Organizations are living, breathing ecosystems composed of individuals unified by a shared purpose and vision. As spiritual practices guide personal growth and self-discovery, these principles can also be applied to an organization's collective journey. The concept of surrender, often associated with spiritual traditions, holds profound relevance for businesses seeking to cultivate adaptability, innovation, and alignment with their core values. By embracing surrender in their operations and strategies, organizations can unlock transformative potential and navigate the complexities of the modern business landscape with wisdom and resilience.

The journey of surrender is a universal human experience that transcends boundaries and belief systems. It is letting go of preconceived notions, embracing uncertainty, and allowing ourselves to be guided by the unfolding of life's plan. Just as individuals undergo transformative personal growth by surrendering to the flow of their experiences, organizations can harness this powerful principle to navigate the complexities of data and AI.

Incorporating Surrender into Business and Data/AI Strategies

To truly integrate the philosophy of surrender into the corporate environment, it is essential to make the concept practical and actionable. **Here are refined examples embodying the principles of surrender that provide a direct correlation to business applications:**

- **Embracing "beginner's mind" and unlearning**:

  - **Business**: This is done by hosting "unlearning workshops" that challenge current assumptions practiced by companies facing technological disruption.

  - **Data/AI**: Apply this to data by conducting "data storytelling workshops" where analysts approach data without biases. This approach is akin to Google's culture of innovation, which encourages employees to think innovatively, creatively, and with open communication (Morein n.d.).

- **Practicing "non-attachment" to outcomes and models**:

  - **Business**: Adopting iterative goal-setting that remains fluid and responsive, similar to how Spotify continuously evolves its music recommendation algorithms based on user behavior and feedback (Hiscox, n.d.). Netflix exemplifies iterative goal-setting by constantly tweaking its user experience.

  - **Data/AI**: Embrace the need for AI models to evolve by setting up systems for continuous monitoring. Learn from platforms like Tesla's Autopilot, which iteratively improves through various means, including software updates (Strutner 2020).

- **Cultivating "trust in the process":**

  - **Business**: Host "reflection circles" to reinforce the reality that challenges lead to growth. At Pixar, trust and creativity are fostered through the Braintrust process, in which trusted colleagues provide direct and constructive feedback to enhance the quality of films (Catmull 2008; Ranadive 2016).

  - **Data/AI**: Hold "retrospective sessions" for AI project teams to share learnings like IBM's approach to AI, where they continually learn from each iteration of Watson's applications (Strutner 2020).

- **Embracing "radical acceptance":**

  - **Business**: Working constructively within present constraints, LEGO embraced its core products and integrated them with popular franchises to revive the brand (Robertson, n.d.). LEGO, facing decline, returned to its brick roots while synergizing creatively with brands like *Star Wars*.

  - **Data/AI**: Accept AI systems' limitations and focus on responsible AI practices, as seen with Fujifilm's diversification from photography to healthcare by leveraging its core technological strengths.

- **Aligning iteration with your why and who**:

  - **Business**: Adopting a "release early, release often" philosophy intertwines strategy (why) with the evolving needs of customers to foster a dynamic dialogue essential to refinement. This approach, exemplified by Intuit collaborating with SmartBear to improve tools like Collaborator based on customer feedback, not only accelerates innovation but deeply aligns product development with the mission and the collective insight of teams (who), ensuring that offerings resonate more profoundly with user expectations (SmartBear n.d.).

  - **Data/AI**: This approach in data/AI can be seen in companies like Google and their products, like Google Data Analytics, which iterates rapidly based on user data. Google continuously collects user feedback and data from its analytics tools and uses this information to make frequent updates and improvements. This iterative approach ensures that their products remain relevant, effective, and user-friendly and that the company can adapt quickly to users' changing needs and behaviors (Culture Partners, 2024).

- **Transparent leadership retreats**:

  - **Business**: These retreats testify to the importance of surrendering control to foster a culture of trust, collaboration, and strategic alignment. By openly discussing strategies, challenges, and opportunities, leaders empower teams (who) to contribute to the organization's mission, ensuring that every member is vested in the journey and its outcomes.

○ **Data/AI**: Tech giants like Salesforce exemplify such transparency, conducting leadership sessions to deliberate on AI's ethical implications and strategize its use for societal benefit (Lee 2021).

- **Fail forward initiatives:**

  ○ **Business**: Embracing setbacks as vital growth steps reflects the why—the purpose driving the pursuit of innovation and resilience. Amazon and Alphabet's X ("moonshot factory") epitomize this culture, treating each failure as a lesson and an opportunity to evolve (HBR IdeaCast n.d.). This approach encourages experimentation and empowers a team to innovate without fear, ensuring that every endeavor, regardless of its immediate outcome, contributes to the organization's collective wisdom and forward momentum.

  ○ **Data/AI**: Similar principles are applied in data science teams at companies like DeepMind, where each project—successful or otherwise—contributes to the collective understanding and development of AI technology (Hillemann, 2023).

By integrating these practices, inspired by the spiritual principles of surrender, organizations can foster a culture of holistic, adaptable, and innovative approaches to business operations and specific data/AI strategies. This mindset promotes agility, resilience, and alignment with the organization's purpose and values, creating a foundation for sustained transformation and growth.

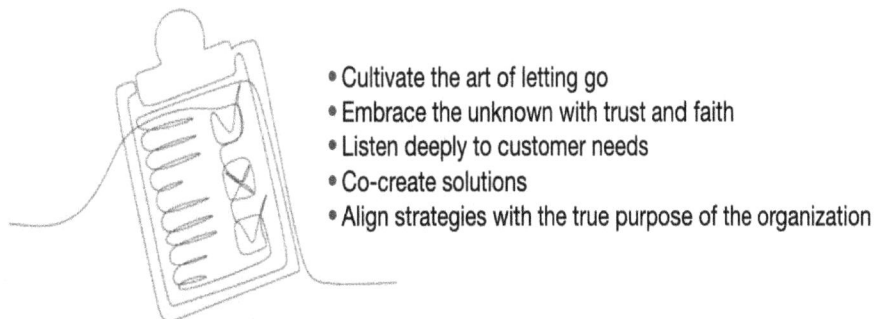

- Cultivate the art of letting go
- Embrace the unknown with trust and faith
- Listen deeply to customer needs
- Co-create solutions
- Align strategies with the true purpose of the organization

Figure 32. Checklist for Mastering Surrender

Case Studies of Surrender in Action

Real-world examples, such as Netflix's data-driven transformation (Daksh 2023) from a DVD rental service to a global streaming titan or Spotify's success with personalized playlists (Rosser n.d.), illustrate the power of surrendering to user insights and adapting business models accordingly. These stories highlight how flexibility, openness to change, and trust in data-driven decision-making can lead to breakthrough innovations and market success. **Here are some other examples:**

- **IBM**: Once a powerhouse in computer hardware, IBM faced a critical turning point in the early 1990s when competition became fierce. Embracing change, it shifted from products like computer chips and printers and focused on software, IT consulting services, and computing research. IBM underwent strategic transformations, including splitting into two companies and embracing AI, to optimize its operations and adapt to market demands. In 2020, IBM furthered its adaptation journey by splitting into two companies. One entity focused on cloud computing and AI, while the other managed IT services. This move was part of IBM's ongoing effort to streamline its operations and concentrate on high-growth areas such as the cloud and AI (Louise 2020; Enderle 2022; Bendor-Samuel 2020).

- **Lego**: The beloved toy company faced near bankruptcy in 2003 due to over-innovation and market saturation with new technology. By recalibrating its approach, Lego embraced a more streamlined business model, revitalized its original products through collaborations with the *Harry Potter* and *Star Wars* franchises, and ventured directly into *Lego* movies. This strategic surrender to market trends and consumer interests allowed the company to rebuild its brand and continue thriving (Strutner 2020).

- **Fujifilm**: As the photography industry shifted to digital, Fujifilm faced decreased demand for its core products. Unlike some competitors, Fujifilm anticipated the digital shift and diversified early. It applied its expertise in film technology to new markets such as pharmaceuticals and cosmetics. This proactive surrender to emerging trends and willingness to reconfigure its business model allowed it to remain successful despite industry upheavals (Guest 2021; Steinlage 2016).

- **Whole Foods and other supermarkets**: During the COVID-19 pandemic, supermarkets like Whole Foods adapted to a surge in online shopping by converting some physical stores into distribution centers to meet the increased demand for deliveries. This pivot demonstrates how surrendering traditional business operations in response to market changes can create new opportunities for efficiency and customer service (World Economic Forum 2021; Wells 2020; Tyko 2020).

These examples show that the concept of surrender, when applied to business, often involves letting go of what once worked to embrace the uncertainty of change and the possibilities of innovation. It is about agilely responding to market conditions and customer needs, often requiring a bold willingness to transform business models and operations.

Ethical Horizons

Surrender also involves a commitment to ethics. The path of ethical exploration is deeply anchored in an organization's core purpose and commitment to achieve (the why) with integrity, respect, and responsibility toward the individuals whose lives technologies touch.

- **The ethics compass (the role of why)**: A commitment to ethical considerations, privacy, and the responsible use of technology reflects the organization's broader mission. It is about striking a balance—leveraging data and AI to drive progress and innovation while rigorously safeguarding against biases, ensuring transparency, and upholding the dignity and rights of every individual. This ethical compass ensures adequate strategies are executed with a deep sense of responsibility and moral integrity.

- **Empowered decision-making (the strength of who)**: The dedicated team members behind ongoing initiatives champion realizing these ethical principles in day-to-day operations and strategic decision-making. Their commitment, insight, and vigilance seamlessly integrate moral considerations into strategies, fostering an environment where responsible leadership is the norm, not the exception. A team's diverse perspectives are invaluable in identifying potential ethical dilemmas and navigating challenges with wisdom and empathy to ensure that advancements in data and AI serve the greater good.

Embracing ethical considerations goes beyond compliance and risk mitigation; it is about aligning technological endeavors with the values and principles that define an organization. Ethical leadership and decision-making are not merely regulatory requirements but fundamental to building trust, fostering innovation, and securing a sustainable future for an organization in the digital age. The why and who are not just guiding lights but the very essence of committing to responsible and ethical transformation.

A Call to Action for Transformative Leadership

By fostering a culture of trust, openness, adaptability, and ethical responsibility, businesses can unlock the full potential of their teams and navigate the challenges of the digital era with wisdom and resilience.

Such an approach illustrates how surrender can guide individuals and organizations toward continuous transformation, driving innovation while maintaining a solid ethical foundation. By embracing the wisdom of letting go and trusting the process, we can unlock new growth frontiers within ourselves and in data and AI.

Integrating the principles of surrender into data and AI strategies can unlock a powerful synergy between offensive and defensive approaches. This harmonious balance fosters innovation and growth and ensures a strong foundation of stability, compliance, and ethical responsibility. Embracing the transformative potential of exploratory data analysis while remaining grounded in the wisdom of surrender, organizations can navigate the complexities of the digital age with resilience, creativity, and a deep commitment to responsible leadership.

Conclusion

The approach presented in this chapter goes beyond traditional data and AI strategy best practices. By integrating core purpose and leveraging the collective team strength, data, and AI can be transformed from mere tools to agents of significant change. A blend of strategic insight and purpose-driven leadership will steer an organization toward success and a legacy of impactful innovation.

Embrace this journey with the knowledge that the strategy, animated by a team's creativity and dedication, is more than a plan—it is a narrative of ambition and transformation. The path forward is illuminated by continuous adaptation, insightful alignment of strategies with organizational values, and the collective wisdom that propels initiatives beyond operational achievements to meaningful growth.

CHAPTER 8

Value Tracking: A Journey of Discernment and Strategic Choice

Value tracking and prioritization are central to the data and AI journey. Strategic imperatives illuminate the route ahead and ensure that each step is taken toward transformative success. Discerning where actual value lies and how to prioritize initiatives becomes a compass for businesses. Through meticulous value tracking and astute prioritization, organizations can make informed decisions that align with their strategic objectives and propel them up the mountain.

Value tracking and dynamic prioritization are often initiated with enthusiasm but, regrettably, sometimes need to be addressed or sporadically updated after that. This can lead to a divergence between initial hypotheses and the evolving realities of data accuracy, technical feasibility, and the ever-changing terrains of technology and consumer behavior. Drawing parallels to the journey of human growth, businesses must adopt growth as their primary KPI. Such a focus ensures that all other KPIs—sales, customer satisfaction, efficiency, and market expansion—are harmoniously aligned to contribute to this central growth objective effectively.

Consider Microsoft under the leadership of Satya Nadella, who has redefined the company's purpose with his vision to "empower every person and every organization on the planet to achieve more." This singular, purpose-driven goal encapsulates the essence of growth, not just in the conventional business sense but as a transformative force for societal and global betterment. Defining their North Star—and continuously tracking value against this overarching guidepost—exemplifies how businesses can dynamically prioritize initiatives to maximize impact and investment in data and AI transformation.

Continuous Value Tracking and Dynamic Prioritization

The essence of success is reaching the summit and appreciating and understanding the steps taken to achieve it. This requires a robust value tracking system that transcends mere numbers, encapsulating quantitative metrics like financial gains and qualitative aspects such as enhanced customer experiences and strategic market positioning. Adopting such a comprehensive system enables organizations to achieve a 360-degree view of their initiatives, leading to an appreciation of the tangible and intangible benefits that contributed to the journey.

Establishing a Continuous Value Tracking System

Reflecting on a significant undertaking with a large oil and gas company of which I was a part, it became evident that focusing on tuning AI models without considering the foundational necessity of high-quality data was a misstep. This scenario occurred during a project aimed at predicting anomalies in oil production, which had the potential to save significant amounts of money, prevent downtimes, and optimize the use of equipment and resources. However, despite the strategic importance of these outcomes, the project should have been prioritized and budget allocated in the first place.

The reason is that strategic alignment is just one dimension of setting priorities. Feasibility, technological viability, data readiness, institutional knowledge, and skill set readiness are also crucial. If we had taken a more holistic approach to prioritization and investment in this use case, we would likely have avoided starting with such a complex approach. Instead, we would have begun with a more straightforward project that still created business value but had more realistic technical and data feasibility, for which the organization was more ready to take on.

This experience underscored the critical need for continuous value tracking and aligning projects with data readiness and technical feasibility criteria to ensure tangible ROI. Just as in personal transformation, in which growth and self-awareness are paramount, business transformations must value growth as the primary KPI. All other metrics are byproducts of this fundamental goal.

Incorporating methodologies like plan-do-check-act (PDCA) and lean principles facilitates a systematic approach to continuous improvement. When applied with a clear understanding of an organization's unique purpose—as exemplified by Microsoft's mission to empower every person and organization to achieve more—these methodologies align initiatives with the overarching goal of growth.

Tools for Effective Value Tracking

- **Kanban boards**: These provide a visual management tool that assists in identifying bottlenecks and areas to prioritize based on current demand, which mirrors how individuals manage personal growth priorities.

- **Five whys and fishbone diagram**: These root-cause analysis tools ensure initiatives target the correct issues. They are akin to introspection practices in personal growth that lead to meaningful insights and improvements.

- **Gemba walks**: Observing and understanding operations firsthand can reveal where value is created and potential improvements, like the personal journey of learning happening through experience.

- **Value stream mapping (VSM)**: This tool aids in identifying inefficiencies and areas for improvement, reflecting the process of evaluating life choices and their alignment with personal goals.

Dynamic Prioritization Based on Continuous Learning

I worked on a retail company's transformation from scattered initiatives to a cohesive strategy. This transformation involved consolidating various independent projects that previously ran in silos. Initially, the company had numerous initiatives targeting different aspects of the business, such as improving supply chain efficiency, enhancing customer experience, and optimizing inventory management. However, these efforts lacked coordination and often competed for resources, leading to inefficiencies and suboptimal results.

Recognizing the need for a more structured approach, we introduced a dynamic prioritization framework. This allowed us to continuously assess and realign projects based on their alignment with the overall strategy, potential value, and feasibility. This approach allowed us to identify which initiatives were most critical to the company's strategic goals and which would likely deliver the highest returns.

For example, we prioritized an initiative to integrate advanced analytics into the inventory management system, which promised quick wins and significant savings. This was chosen over a more ambitious project to develop a new customer engagement platform, which—although strategically important—required a longer timeline and substantial upfront investment in data infrastructure and skills development.

Due to the chosen process, the company transitioned from a fragmented approach to a cohesive strategy aligned with its long-term objectives. This mirrors a personal journey where alignment with excitement, joy, and ease—our internal markers for growth and fulfillment—guides decision-making. Dynamic prioritization has the power to align with overall strategy and continuous value tracking. Just as personal development is driven by aligning with what feels right internally, organizational growth is driven by aligning initiatives with strategic goals and continuously reassessing their impact and feasibility.

Successful scaling of AI projects often involves a portfolio approach, which considers a mix of high-impact, feasible initiatives with long-term strategic goals. This method ensures that the most promising projects get the support they need while maintaining alignment with the overarching business strategy .

Lean Six Sigma: A Balanced Approach

Lean Six Sigma, a team-focused managerial approach, seeks to improve performance by eliminating resource waste and defects. Combining Six Sigma methods and tools with the lean manufacturing philosophy, it strives to eliminate the wasting of physical resources, time, effort, and talent while assuring quality in production and organizational processes (Kenton, 2023).

Originating from lean principles established by Toyota in the 1940s and the Six Sigma strategic approach developed by Motorola in the 1980s, Lean Six Sigma emphasizes reducing inefficiencies and enhancing quality. The lean aspect focuses on eliminating any use of resources that does not create value for the end customer. At the same time, the Six Sigma targets improve output quality by reducing defects and variability in processes. Together, they provide a comprehensive strategy for performance improvement.

For instance, Lean Six Sigma projects often use the define, measure, analyze, improve, and control (DMAIC) cycle, a data-driven method for optimizing business processes. This methodology involves defining the problem, measuring key aspects, analyzing data, implementing improvements, and establishing controls to sustain gains. Techniques such as Kanban for workflow management, Kaizen for continuous improvement, and value stream mapping for waste elimination are integral to Lean Six Sigma practices.

Incorporating Lean Six Sigma principles can substantially improve both business and personal growth. Companies can achieve higher efficiency and better outcomes by eliminating waste and enhancing quality. Similarly, on a personal level, adopting such principles can help individuals streamline their activities, reduce stress, and improve overall effectiveness. Just as businesses benefit from a balanced approach to efficiency and quality, individuals can achieve personal growth by aligning their actions with these principles.

Integrating Growth as the Primary KPI

Growth is the only KPI that genuinely matters—all other metrics are its byproducts. A singular focus on growth allows for a more precise evaluation of initiatives and how they align with an organization's unique purpose. Just as individuals gauge alignment with their core values through feelings of excitement, joy, and ease, businesses can measure their alignment through growth and the seamless integration of initiatives. By adopting this perspective, organizations can transcend competing KPIs and ensure that short-term decisions and long-term strategies harmoniously align with the mission.

Case Study: Amazon

Amazon stands out as a quintessential example of leveraging internal value tracking to steer both its short-term actions and long-term strategic direction. It developed methodologies for prioritizing investments and initiatives, underpinning its reputation as a customer-centric and innovative powerhouse (Galea-Pace 2020).

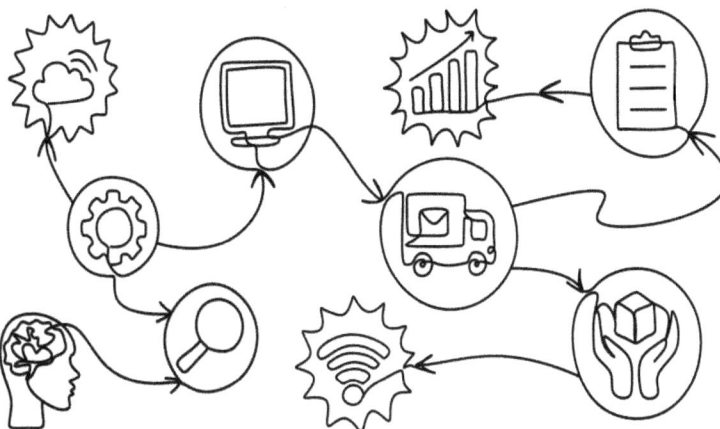

Figure 33. Optimizing Supply Chains through Data Analytics and AI Decision-Making

At the heart of Amazon's success is its adept use of real-time analytics to enhance customer experiences and streamline its supply chain, ensuring customer needs and market trends directly influence its service offerings. A pivotal element in Amazon's data and AI prowess is Amazon Web Services (AWS), which provides a robust infrastructure for large-scale data analysis and AI initiatives.

AWS's capabilities allow Amazon to analyze vast amounts of data and innovate rapidly, developing new services that meet evolving customer expectations. This holistic approach to value tracking and prioritization showcases how deep insights into customer behavior and preferences can drive strategic decision-making, leading to sustained market leadership and innovation.

- **Continuous data-driven decision-making**: An unwavering customer obsession is at the core of Amazon's strategy, guiding every facet of its operations and strategic decisions. An extensive data analytics framework powers the company's dedication to understanding and anticipating customer needs. This framework captures customer interactions and transforms this data into actionable insights, driving product and service innovations that align with customer expectations and emerging trends.

- **Dynamic project prioritization**: Amazon's approach to project prioritization is anything but static. By harnessing detailed analytics across various metrics—ranging from customer satisfaction scores to operational efficiency indicators—Amazon ensures a flexible prioritization process. This process adeptly identifies projects that align with key strategic goals, such as enhancing the customer experience or achieving operational excellence, ensuring optimal resource allocation and focus.

- **Iterative development with feedback loops**: The hallmark of Amazon's development philosophy is its iterative nature, which is significantly informed by direct customer feedback. Through real-time data collection methods, such as A/B testing and customer reviews, Amazon maintains a continuous improvement cycle. This cycle allows for the agile refinement of products and services, ensuring customer feedback directly influences development trajectories and investment decisions.

- **Long-term investment philosophy**: Amazon's strategic outlook is characterized by a profound commitment to long-term value creation. This principle has driven some of its most impactful investments, including the development of AWS. Through sophisticated internal tracking and analysis, Amazon identifies initiatives with the potential to redefine markets or create new ones, embracing the necessary patience and investment to cultivate these opportunities until they achieve market leadership.

- **A comprehensive approach to measuring ROI**: Unique in its approach, Amazon evaluates the ROI of its initiatives through a lens that extends well beyond traditional financial metrics. Amazon ensures a nuanced assessment of investment returns by considering the synergistic effects between products and services, customer lifetime value, and strategic alignment. This comprehensive ROI perspective supports informed decision-making, balancing immediate outcomes with strategic, long-term objectives.

Amazon exemplifies how data-driven insights and strategic foresight can harmonize to guide a company toward sustained innovation and market leadership by meticulously applying internal value tracking and an advanced understanding of ROI. It is a potent combination of customer-centricity, agile development, and strategic investment driving organizational success in the digital age.

Evaluating Impact and Alignment

Every data and AI initiative should be aligned with the organization's strategic objectives and vision. Just as climbers prioritize their actions based on their progress and alignment with the summit vision, organizations must evaluate the impact and alignment of each initiative to stay on the right path.

Through continuous evaluation, organizations can ensure that their data and AI initiatives contribute to realizing transformative goals. This enables course corrections, realignments, or even discontinuing projects that no longer serve the greater purpose. It allows a focus on initiatives that drive value and align with long-term objectives.

Case Study: Zara

Zara, a flagship brand of the Spanish retail group Inditex, stands out for its unique business model that integrates data and AI to stay aligned with rapidly changing fashion trends and consumer preferences

(Uberoi 2017). Unlike Blockbuster, which failed to adapt its business model in the face of changing industry dynamics, Zara's agility and data-driven strategy have been critical to its success in the highly competitive fast-fashion industry.

This is another compelling case study in strategic value tracking and prioritization. Success hinges on a responsive supply chain and advanced trend analysis capabilities, allowing for the rapid turnover of trend-sensitive product lines underscored by its use of real-time data and analytics to adjust operations and inventory in response to current fashion trends and consumer demands. This ensures that Zara meets and often anticipates the market's evolving preferences.

Zara utilizes real-time data from its stores worldwide to track customer preferences and fashion trends as they emerge. This data-driven approach enables Zara to quickly design, produce, and distribute clothing that meets current consumer demands, significantly reducing the fashion industry's traditional lead times. This allows Zara to discontinue underperforming items and introduce new designs that are more likely to resonate with customers.

Employing AI algorithms, Zara optimizes its supply chain and inventory management. By analyzing sales data and stock levels in real-time, the system predicts future demand more accurately, ensuring that popular items are quickly restocked and underperforming stock is reduced. This makes desired items more readily available, which enhances operational efficiency and supports its commitment to sustainability by reducing overproduction and waste.

Zara's success story highlights the importance of aligning data and AI initiatives with an organization's strategic objectives and the power of continuous evaluation to stay ahead in a dynamic market environment. Their approach illustrates the critical role of timely data-driven decisions in achieving operational excellence and market responsiveness, reinforcing the importance of aligning AI and data initiatives with strategic business objectives.

Quantitative and Qualitative Benefits

Understanding the full spectrum of quantitative and qualitative benefits is essential to appreciate the transformative power of effective value tracking and prioritization. These practices contribute to measurable outcomes, such as increased profitability and market share and enhance aspects of business operations that, while harder to quantify, are equally vital for long-term success.

- **Quantitative benefits**:
  - **Increased profitability**: Implementing data-driven strategies through meticulous value tracking enables organizations to optimize operational efficiency, reduce costs, and enhance revenue generation. This strategic approach leads to significant improvements in profitability, demonstrating the tangible financial gains from well-informed decision-making processes.

- **Market share growth**: Data and AI-driven prioritization allow businesses to swiftly identify and act on market opportunities, resulting in expanded customer bases and increased market share. By staying ahead of trends and adapting to consumer demands with agility, companies can secure a competitive edge, attract customers, and capture more significant market segments.

- **Enhanced customer satisfaction**: Aligning product development and service offerings with customer needs—identified through comprehensive value tracking—directly contributes to higher customer satisfaction. Satisfied customers are more likely to engage in repeat business, recommend services to others, and contribute positively to brand loyalty, further driving the organization's success.

- **Qualitative benefits**:

  - **Improved brand reputation**: Strategic initiatives grounded in value tracking meet and often exceed customer expectations, fostering a positive brand image. This strengthens the company's standing in the market, attracting new customers and retaining existing ones through perceived value and trustworthiness.

  - **Employee engagement and innovation**: A culture that embraces data-driven decisions and continuous improvement encourages employee engagement by involving team members in the company's strategic direction. This fosters a sense of ownership and pride, leading to higher creativity, innovation, and overall job satisfaction.

  - **Adaptability and resilience**: The qualitative benefit of organizational agility, manifested through the ability to swiftly pivot strategies based on data insights, positions a company as a resilient leader capable of thriving in dynamic market conditions. This ensures long-term viability and success, as the organization can quickly respond to challenges and seize new opportunities.

The synergistic relationship between quantitative and qualitative benefits underlines the comprehensive value of effective tracking and prioritization. While financial metrics are crucial, the broader impacts on brand, culture, and market position highlight the holistic advantages of these practices. Together, they build a robust, forward-thinking organization poised for continuous growth and innovation.

Leadership and Company Culture

These are pivotal in successfully implementing value tracking and prioritization. The direction set by leaders and the culture fostered within an organization significantly influence the ability to adapt, innovate, and transform.

- **The impact of leadership**:

  - **Vision**: Leaders who articulate a vision for integrating data and AI transformation goals inspire and motivate the organization when they demonstrate a commitment to these objectives. Modeling the behavior they wish to see throughout the organization sends a powerful message.

o **Strategic decision-making**: Prioritizing value tracking and data-driven insights in the decision-making processes sets a precedent for the entire organization. A leader's approach to making informed strategic decisions influences the prioritization of projects and the allocation of resources, ensuring alignment with long-term goals.

o **Fostering a culture of innovation**: Leadership is crucial to creating an environment that encourages experimentation, learning from failures, and celebrating successes. Such a culture is essential for continuous improvement and the iterative development of data and AI initiatives.

- **The influence of organizational culture**:

  o **Reliance on data**: An organizational culture that values evidence over intuition encourages employees at all levels to engage with data and analytics. This cultural shift is necessary for tracking and prioritizing value in data and AI initiatives.

  o **Transparency and open communication**: Cultivating these as organizational norms ensures that insight and learning from value tracking is shared across the institution. Such openness fosters collaboration and collective action towards common goals.

  o **Adaptability and knowledge**: A culture that promotes elasticity and continuous learning empowers the organization to respond swiftly to insights gained from value tracking. This encourages teams to pivot or realign strategies based on data-driven evidence, which enhances the organization's agility and resilience.

Leadership and culture are not just peripheral factors but central elements that directly impact the success of value-tracking and prioritization initiatives. Leaders committed to a data-driven vision and cultivating a culture that embraces these principles can significantly enhance their organization's ability to navigate the complexities of transformation successfully.

Making Informed Decisions

Every step in a climb is a decision. In the data and AI landscape, accumulating decisions are pivotal to the long-term success of the transformation journey. With a well-established value tracking system, organizations have the insights needed for data-driven decision-making. Projects can be prioritized based on strategic significance, ensuring that every step taken is in the direction of maximizing ROI. Just as climbers rely on their knowledge and experience to make vital decisions while moving up the mountain, organizations must base their choices on data-driven insights.

The value tracking system provides organizations the information needed to make informed decisions. It helps leaders and stakeholders understand each initiative's benefits and potential risks, enabling them to prioritize projects better.

Case Studies: Leveraging Data-Driven Decision-Making

In the current business landscape, integrating data-driven decision-making processes is a pivotal strategy for organizations aiming to enhance operational efficiency, improve customer experiences, and drive innovation. This subsection delves into how renowned companies like McDonald's, Netflix, Google, and Starbucks have succeeded.

- **McDonald's**: Acquiring Dynamic Yield in 2019 marked a significant step towards harnessing big data for personalized customer experiences. By leveraging algorithmically-driven decision-logic technology, McDonald's aimed to personalize menu displays at drive-throughs based on weather, time of day, and trending menu items. This strategic move sought to enhance customer satisfaction and optimize sales by catering more directly to consumer preferences and situational demands (Unscrambl 2021).

- **Netflix**: Its utilization of data analytics exemplifies the power of data in creating and recommending content that resonates with viewers' preferences. By analyzing extensive data sets—including subscriber ratings, search patterns, and viewing habits—Netflix has successfully predicted and produced widely acclaimed series such as *House of Cards* and *Arrested Development*. This approach underscores the value of data-driven insights in content curation, significantly contributing to Netflix's reputation for compelling and relevant programming.

- **Google**: Project Oxygen is a testament to the application of data analytics in leadership development and employee satisfaction. By analyzing performance reviews and feedback surveys, Google identified the critical behaviors of high-performing managers, which informed the development of targeted training programs. These initiatives have improved managerial effectiveness, showcasing the impact of data-driven strategies on organizational culture and leadership quality (Unscrambl 2021).

- **Starbucks**: Taking an analytical approach to selecting store locations demonstrates the utility of data in optimizing operational decisions. Partnering with location analytics companies, Starbucks assesses demographic and traffic pattern data to identify ideal locations for new stores. This data-driven strategy has enabled Starbucks to make informed decisions about new investments, significantly enhancing the likelihood of success for each new venture (Unscrambl 2021).

These cases illustrate the transformative potential of data analytics across various aspects of business operations, which has proven crucial in achieving business objectives and maintaining competitive advantage. Organizations looking to leverage data-driven strategies can focus on sourcing relevant data, building predictive models, and undertaking necessary organizational transformations to ensure that data insights translate into actionable, impactful decisions.

As evidenced by the success stories of these leading companies, a strategic and integrated approach to data analytics can yield significant improvements in decision-making and overall business performance (Stobierski 2019).

Maximizing Return on Investment

Organizations strive to optimize their resources and investments as climbers seek to maximize their efforts by choosing the best reward-versus-risk paths. Portfolio value tracking and prioritization enable organizations to allocate resources strategically, ensuring that high-value projects receive the necessary attention and investment (Datacenters, 2023). Organizations can enhance their ROI and accelerate their data and AI transformation journey by focusing on initiatives that deliver the most significant value. This approach empowers the achievement of meaningful outcomes while managing resources, time, and efforts effectively (Convolo 2024).

A notable illustration of this approach is highlighted by Performance Improvement Partners, which uses data-driven decision-making to enhance investment strategies and outcomes. Adopting data analytics in private equity involves shifting from traditional instinct-based decision-making to a more scientific and data-oriented approach. This transition is crucial for navigating the complexities of the modern market, enabling firms to leverage existing data assets into actionable insights. According to a 2020 Deloitte report, this data-driven mindset is imperative for leaders seeking to adapt to a rapidly changing environment (Fellows, 2023).

Data analytics afford private equity professionals a managed and measured approach to investment, uncovering opportunities that might otherwise remain obscured within internal data. This systematic strategy identifies potential investments and assesses their viability and profitability more precisely (Legan 2023).

For private equity firms, the application of data analytics spans the entire investment lifecycle. From pre-acquisition analysis (forecasting revenue and assessing cost structures) to post-acquisition growth strategies, data analytics provide the insights necessary for delivering ROI and maximizing returns. Analyzing a portfolio company's data assets can reveal critical areas for improvement, such as underperforming products or inefficient business processes, guiding strategic interventions for enhanced profitability (Ostili, 2024).

Key takeaways:

- **Strategic alignment**: Aligning data-driven insights with investment strategies ensures that resources are allocated to high-potential opportunities (AIContentfy 2024).

- **Competitive advantage**: Leveraging data analytics provides a distinct edge over competitors, enabling firms to identify and capitalize on investment opportunities more effectively (Pappas, 2023).

- **Enhanced decision-making**: A data-oriented approach to investments enhances accuracy and reduces risks associated with market dynamics.

Case Study: Blackstone

By integrating data-driven insights into its strategic planning, Blackstone achieves superior investment returns and sets a benchmark for operational excellence within the industry. It leverages advanced analytics to inform investment decisions, enhance operational efficiencies, and identify growth opportunities across its portfolio. This approach underlines the critical role of comprehensive data analysis in enhancing competitive advantage, driving value creation, and facilitating informed decision-making, exemplifying that significant ROI can be realized through the strategic application of data analytics in private equity (Parati, 2023).

The strategic utilization of data analytics in private equity exemplifies a broader trend across industries toward leveraging data for informed decision-making and ROI maximization.

Integrating Spiritual Wisdom

The objective and critical results (OKR) framework is a widely respected tool designed to help organizations align and measure goals against tangible outcomes. Devised by Andy Grove (the former CEO of Intel) and later popularized by John Doerr (a renowned venture capitalist), OKRs have been instrumental in driving the success of major tech companies like Google, LinkedIn, and Twitter. The framework is applicable across fields beyond technology, including healthcare, education, and the nonprofit sector, demonstrating its versatility and effectiveness in ensuring strategic alignment and fostering growth.

OKRs consist of setting a qualitative objective—a broad goal designed to be inspirational—and attaching to it a set of quantifiable vital results and specific, time-bound achievements that can be measured to gauge progress toward the objective. This framework encourages transparency, alignment, and engagement across all levels of an organization.

Integrating spiritual wisdom through OKRs involves setting goals that advance technical proficiency and align with the organization's larger vision and ethical considerations. This holistic approach ensures that technological advancements contribute positively to the organization's mission and societal well-being.

To successfully implement OKRs, it's essential to:

- **Set clear objectives**: These should be ambitious yet achievable, motivating teams to strive for excellence, such as improving customer experience through AI-driven personalization (Datacenters, 2023).

- **Define measurable key results**: These should quantify the achievement of each objective, such as enhancing user engagement by a specific percentage within a defined period (Convolo 2024).

- **Ensure alignment across teams**: OKRs at the organizational level should cascade to every team and individual, promoting unity and focused effort toward common goals (Fellows 2023).

- **Regular check-ins**: Frequent progress reviews against OKRs help identify obstacles and adjust strategies as needed (Legan, 2023).

- **Maintain flexibility**: The dynamic nature of data and AI projects requires adaptability in goal setting and execution (Ostili 2024).

- **Celebrate and reflect**: Acknowledging successes and learning from setbacks fosters a culture of continuous improvement and innovation (AIContentfy 2024).

OKRs' credibility lies in their proven track record of driving growth and innovation by aligning teams around focused and measurable goals. This framework enhances operational efficiency and promotes a culture of accountability and continuous learning.

OKR Examples

- **Objective**: Improve the accuracy of an AI-driven customer support chatbot.

  - **Key result 1**: Reduce the number of escalated chatbot interactions with human agents by 25 percent in the next quarter.

  - **Key result 2**: Achieve a 90 percent positive feedback rate on chatbot interactions by the end of the year.

- **Objective**: Enhance an e-commerce platform's product recommendation engine.

  - **Key result 1**: Increase the click-through rate of recommended products by 30 percent in the next six months.

  - **Key result 2**: Achieve a 15 percent increase in sales from recommended products in the next quarter.

- **Objective**: Optimize supply chain operations using AI-driven forecasting.

  - **Key result 1**: Achieve a 95 percent accuracy rate in demand forecasting for the next quarter.

  - **Key result 2**: Reduce stockouts by 20 percent and overstock by 15 percent in the next six months.

Case Study: Google's Use of OKRs

While there is no direct evidence that Google DeepMind specifically used OKRs to drive AI advancements, Google has widely applied the OKR framework to achieve various objectives, including organizational alignment and fostering innovation. One notable achievement, driven by DeepMind's AI algorithms, was a 40 percent reduction in data center cooling energy consumption (Google 2016; Wired 2016), leading to a 15 percent overall reduction in total energy usage (The Guardian 2016).

By integrating the spiritual wisdom of setting meaningful objectives with the tactical precision of OKRs, organizations can ensure that data and AI initiatives are impactful and aligned with their broader mission. As a world-class AI transformation expert, I emphasize the importance of using OKRs to drive value, alignment, and success. By defining clear objectives, measuring key results, and fostering a culture of transparency and collaboration, organizations can effectively navigate the complexities of their transformational journey and achieve strategic goals with confidence and purpose.

Incorporating Foundational Values and Mission

Drawing insights from *The Buddha and the Badass: The Secret Spiritual Art of Succeeding at Work* by Vishen Lakhiani, it becomes evident that identifying core values is pivotal. An effective exercise to uncover these is reflecting on one's origin story—considering essential life experiences that shape beliefs and principles.

This reflective practice is a journey into the past and a compass for future decision-making and goal-setting. Grounding projects and initiatives in foundational values ensure a resonant and purposeful path forward.

- **The three most important questions (3MIQ)**: Lakhiani introduces this framework to swiftly align team motivations with organizational goals. Spending just a few minutes identifying top experiences, growth areas, and contributions can unveil profound insights into what drives and fulfills a team. Implementing the 3MIQ framework fosters a work environment where every initiative advances technical objectives and feeds into team members' personal growth and satisfaction.

- **The power of communicating the why**: In an era where engagement and authenticity are paramount, articulating the "why" behind work is crucial. Beyond delineating tasks and objectives, inspiring a team with a clear and compelling purpose can catalyze motivation and dedication. This entails outlining goals and weaving the narrative of how each initiative contributes to the broader mission to make a meaningful impact—whether innovating in AI, transforming data management, or driving sustainable growth.

- **Embracing stealth leadership**: *The Buddha and the Badass* advocate for "stealth leadership," a principle where titles do not confine leadership but are an attribute anyone can embody in the organization. This perspective encourages every team member to initiate change, lead by example, and contribute ideas that align with the organization's values and mission. It is a call to action for individuals at all levels to engage proactively with their work, championing initiatives that resonate with their values and the organization's strategic goals.

Integrating foundational values into goal setting and prioritization is not merely a strategic advantage but a necessity for meaningful success in today's complex digital landscape. Organizations can foster an innovative, thriving, deeply fulfilling culture aligned with a greater purpose by employing reflective practices, leveraging frameworks like 3MIQ, communicating purposefully, and encouraging leadership at all levels.

Such a holistic approach ensures that the journey of data and AI transformation is navigated with clarity, purpose, and a profound sense of fulfillment, transforming challenges into opportunities for growth, innovation, and impactful contributions to society.

Addressing Real-World Challenges

Value tracking and prioritization ensure that initiatives align with strategic goals and produce tangible results by addressing challenges directly. This proactive approach and the clear understanding that flows from it enable organizations to turn challenges into opportunities.

- **Define clear objectives**: Setting straightforward and meaningful objectives that align with an organization's vision and strategic goals is crucial. Objectives should be ambitious yet attainable, inspiring teams to work towards a common purpose.

- **Measure key results**: Identifying and evaluating specific, quantifiable outcomes that indicate progress toward objectives is essential. Clear measurement criteria help gauge success and areas needing improvement.

- **Balance short-term and long-term goals**: It is challenging to balance short-term wins and long-term strategic objectives. Prioritizing and allocating resources wisely is vital to achieving both immediate value and transformative initiatives.

- **Alignment and transparency**: Ensuring coordination across teams and departments is vital for successful value tracking and prioritization. Clear communication of objectives and results fosters collaboration, avoids duplication, and enhances organizational alignment.

- **Data availability and quality**: Reliable information is fundamental for effective value tracking. Organizations must maintain accessible, high-quality data through robust governance and management practices.

Consider a scenario where a retail company implements AI chatbots to improve customer service. Despite training them extensively, real-world interactions reveal nuances and complexities the system needs to address initially. This highlights the importance of continuous learning and adaptability to enhance the customer experience and drive business value.

A holistic approach in addressing challenges in data and AI transformation not only ensures alignment with strategic goals but also emphasizes the critical link between overcoming obstacles and effectively tracking and prioritizing the value of data and AI initiatives.

Value Tracking and Prioritization Best Practices

Successful mountain climbing expeditions are based on the best practices followed by the climbers. Similarly, incorporating value tracking and prioritization into data and AI transformation involves adopting best practices that ensure strategic objectives align with actionable insights.

- **Top-down and bottom-up approaches**: Organizations should adopt a top-down approach to set high-level strategic objectives while also encouraging a bottom-up approach for teams to contribute insights and suggest key results that align with the overall strategy.

- **Regular check-ins and updates**: Continually gauging operations with progress updates helps teams stay on track and adjust as needed. Check-ins allow discussion of challenges, sharing learnings, and ensuring everyone is focused on achieving the objectives.

- **Flexibility and adaptability**: Data and AI transformation is dynamic; priorities may shift as new insights are gained or market conditions change. Organizations should be flexible and adaptable and adjust objectives and key results when necessary.

- **Learn from failures**: Not every initiative will succeed, which is okay. Organizations should foster a culture encouraging a greater understanding of failures and use such lessons to inform future decision-making.

- **Continuous improvement**: Value tracking and prioritization should be iterative. Organizations should continuously evaluate the effectiveness of their OKRs, gather feedback, and improve their approach over time.

- **Collaborative decision-making**: Encourage a culture of cooperative decision-making involving all relevant stakeholders. This ensures that different perspectives are considered, leading to more informed and holistic choices. For instance, when deciding on a new data initiative, involve technical teams and business units to ensure alignment and feasibility.

- **Data literacy training**: By ensuring that employees at all levels understand the basics of data and its significance, a culture where data-driven decision-making becomes the norm is fostered. Regular workshops, seminars, or online courses can enhance data literacy.

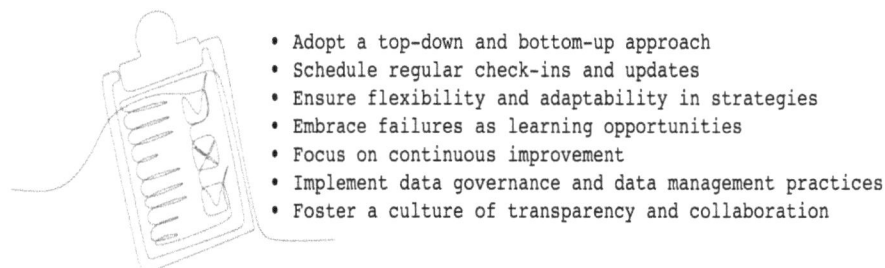

- Adopt a top-down and bottom-up approach
- Schedule regular check-ins and updates
- Ensure flexibility and adaptability in strategies
- Embrace failures as learning opportunities
- Focus on continuous improvement
- Implement data governance and data management practices
- Foster a culture of transparency and collaboration

Figure 34. Checklist for Implementing Best Practices

These best practices can further enhance organizations' data and AI transformation approaches, ensuring a more comprehensive and effective strategy.

Case Studies

Airbnb's decision-making exemplifies a hybrid approach, blending top-down goals with bottom-up feedback. Leaders define vital objectives, and team insights influence implementation. This was apparent when they revamped their digital platforms, incorporating strategic aims with user input to refine functionality and design. This detailed, collaborative strategy is vital to tailoring services that meet business objectives and customer needs, showcasing a model for others aiming to balance innovation with user-centered design (Yip 2017).

Likewise, Microsoft's transformation is a profound example of how an organization can realign its mission and operational approach to foster significant change. Under Satya Nadella's leadership, Microsoft adopted a growth mindset that was not just about expanding its product line and reshaping company culture. Microsoft realigned its values by prioritizing innovation and committing to empowering every individual and organization, driving a holistic transformation. This strategy was about tracking progress and ensuring that every initiative contributed to the company's broader vision, resulting in improved market position and cultural revitalization. For a deeper understanding of Microsoft's transformation, explore its business case study repositories and Microsoft's official corporate communications (Ibarra & Rattan 2016).

Conclusion

Ensuring that every initiative is carefully planned and aligned with the overriding vision is akin to a climber selecting the best path up a mountain. Like a mountaineer who appreciates each step's value, businesses must measure and embrace both the tangible and intangible benefits of data and AI endeavors. With these principles as a guide, leaders will be equipped to advance confidently, transforming challenges into opportunities for growth and ensuring an impactful journey aligned with the broader mission.

CHAPTER 9

Architecting and Implementing Data-Driven Technology Ecosystems

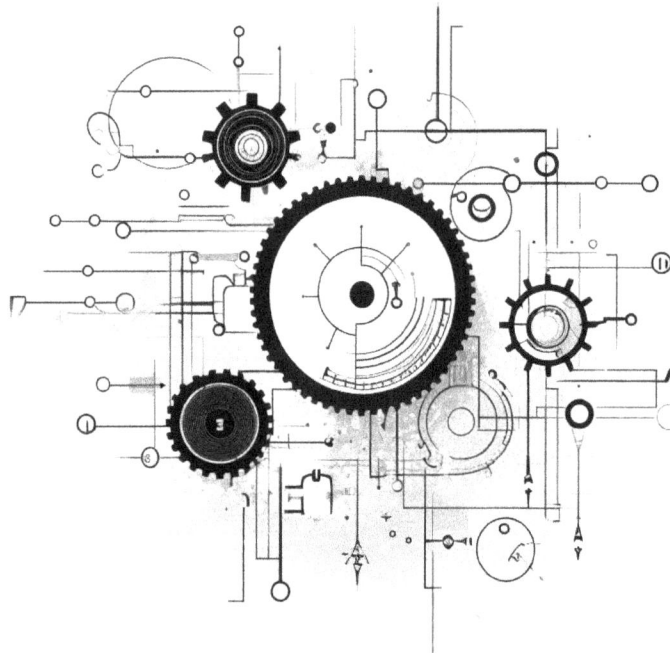

Having covered the strategic imperatives of value tracking and prioritization, it is time to cross to the concrete and engage in the transformative process of constructing and implementing a technology ecosystem tailored for data and AI success. The transition from vision to value is both a strategic endeavor and a tactical challenge, demanding more than a clear vision or the ability to prioritize. What is required is the careful assembly of a robust technology infrastructure—which means navigating a complex maze of tools and platforms, each promising to unlock the potential of data and AI.

Marrying the strategic with the tactical, this chapter will serve as a comprehensive guide that bridges the gap between understanding the critical importance of aligning data and AI initiatives with core business objectives and navigating the practicalities of bringing this alignment to fruition. I will cover ecosystem design and implementation, emphasizing the need to focus on outcomes over technology, building for agility, and managing diversity within the tech stack. These principles can guide organizations through the plethora of choices, ensuring that each decision is made with strategic intent and is harmoniously aligned with growth objectives and ethical considerations.

However, this chapter is not only about constructing a supportive technology infrastructure but also an invitation to traverse the delicate balance between visionary aspirations and tangible value. The goal is to ensure that technology investments are strategic and reflect deeper values and aspirations—including moving towards a future where technology amplifies human potential, enabling growth, innovation, and sustainable, ethical transformation centered on the well-being of all.

Principles for Modern Data and AI Architecture

The architectural principles for data and AI success—such as scalability, flexibility, security, robustness, interoperability, and maintainability—become not just technical guidelines but reflections of strategic imperatives. Each principle is a way to carefully balance technological prowess and strategic foresight to ensure that built tech ecosystems meet today's challenges and evolve with tomorrow's advances.

I will now present several examples of companies that undertook projects showcasing these principles.

Scalability: Growing with Grace

Netflix's journey from a DVD rental service to a global streaming giant exemplifies scalability. By successfully transitioning to managing over 500 cloud-based microservices, it handled nearly 2 billion API requests daily by 2015, significantly reducing costs and improving service quality (Krishna 2021; Varshneya 2021).

Initially, Netflix's infrastructure supported a simple business model; however, as they transitioned to streaming, they faced the monumental task of delivering vast amounts of content reliably across the globe. Netflix adopted a microservices architecture, allowing it to scale resources dynamically as demand fluctuated. This approach accommodated growth and enabled rapid innovation, allowing Netflix to introduce new features and continually improve the user experience. Their scalable architecture is a testament to how technological foresight can facilitate growth and transformation, aligning perfectly with strategic ambitions to dominate the global entertainment landscape (Mauro 2015; Krishna 2021; Varshneya 2021).

Flexibility: Embracing AI-Driven Change

Tesla's approach to continuous improvement in its Autopilot features showcases the principle of flexibility. Through over-the-air software updates, Tesla has an unparalleled ability to adapt and enhance its autonomous driving features. This flexibility stems from a culture of innovation, supported by a robust technological infrastructure that allows AI algorithms to enhance driver assistance features and improve safety. The product evolves with users, reflecting a deep alignment between Tesla's technological strategies and its mission to accelerate the world's transition to sustainable energy (Shah, 2023; Alexander, 2023; Q.ai, 2022).

Security, Ethical AI, and Data Protection

Google's implementation of differential privacy in its AI models, such as those predicting traffic patterns in Google Maps, illustrates the critical importance of security and ethical AI. Google ensures individual user data remains anonymous while providing valuable insights. This approach reflects a broader commitment to respecting user privacy and maintaining trust, a crucial component of Google's business strategy in an era where data security and privacy concerns are paramount (Google for Developers 2021; Google Privacy & Terms n.d.).

Robust Data and AI Architectures in Fraud Detection

In the financial sector, robust data and AI architectures are critical to enhancing security measures, particularly in fraud detection systems utilized by banks and financial institutions. Companies like PayPal have developed advanced AI algorithms that continuously monitor transactions to detect and prevent fraudulent activities. These systems demonstrate robustness by swiftly identifying anomalies and suspicious patterns in real time despite dealing with a vast amount of financial data and ever-evolving fraud tactics.

By leveraging ML and AI technologies, these fraud detection systems can adapt to new fraud schemes, learn from past incidents, and enhance their accuracy over time. This resilience is essential for safeguarding financial transactions and protecting customers against potential security threats. The ability of these AI architectures to maintain reliability and effectiveness in detecting fraudulent activities highlights the importance of robust design principles in fortifying data-driven systems within critical sectors such as finance (Cloud 7 IT Service 2023; Architech 2024; Owen 2022).

Interoperability: Seamless AI Integration

Adobe's Experience Platform demonstrates interoperability by integrating AI models and data across Adobe's suite of products and third-party systems. This seamless integration enables marketers to deliver personalized customer experiences via different channels, leveraging AI to analyze and act on data from various sources. Adobe's focus on interoperability supports its strategic goal of empowering creatives and marketers with comprehensive and cohesive tools, highlighting the business value of a unified technology ecosystem.

According to a post on Adobe's blog, a partnership between Adobe and Databricks enables brands to compose data- and AI-driven insights to create personalized customer experiences at scale and speed. This is achieved through enhanced integrations between the Adobe Experience Platform (AEP) and Databricks, which enables brands to reduce data movement, govern the use of sensitive data while maintaining interoperability, and minimize friction in finding insights with machine learning (Garfield, 2024). A Databricks' blog post also confirms that the strategic partnership allows for seamless integration of predictive models into the AEP that deliver hyper-personalized experiences at scale (Sobel 2024). Adobe has also announced major Adobe Experience Cloud innovations that enable brands to unify customer data across their organization, a necessary ingredient to implementing and deriving value from generative AI (Adobe 2024).

A *Computerworld* article mentions that Adobe's AI assistant offers predictive insights and recommendations based on customer data, made possible by the generative experience model that comprises base models trained on data such as product information, community forums, best practices, and custom models AEP customers can opt into (Finnegan 2024). Another Adobe blog post introduces Adobe Experience Platform AI Assistant, which can answer a wide variety of questions, automate tedious tasks, simulate outcomes, and generate desired audiences and journeys using simple English language prompts (Bhambhri 2024), while a GeniusOS article confirms that Adobe has unveiled it for customer experience management (quade 2024). An article published by One North also highlights AI within the AEP, enabling brands to deliver personalized and seamless experiences at scale (Lill 2023).

Maintainability: Evolving with AI Advances

Nvidia's software stack (including CUDA and cuDNN) is designed for maintainability, ensuring compatibility across generations of AI hardware. This foresight allows developers to leverage the latest graphics processing unit (GPU) technology advancements, facilitating continuous improvement in AI model performance without significant overhauls. Nvidia's commitment to maintainability and backward compatibility exemplifies how technology can be architected not just for current needs but with an eye toward future evolution, aligning with the company's strategic vision of driving breakthroughs in AI and computing (Saunders, Vaidya, Tetelman, & Spirin 2024; CIO Influence 2023).

Activity:

- **Crafting your ecosystem with scalability in mind**: Understanding the scalability demonstrated by Netflix, please consider how your organization can prepare for growth and innovation. Start by evaluating your current infrastructure's ability to handle increased loads and the ease with which new services or features can be integrated. *Ask yourself,* "How can our technology infrastructure adapt to sudden spikes in demand?" and "Are we prepared to incorporate new functionalities without disrupting existing services?"

- **Embracing flexibility for future-proofing**: Inspired by Tesla's adaptability, reflect on your organization's readiness to adopt new technologies or methodologies. *Ask yourself,* "Do we have a process for integrating new technologies or data sources with minimal disruption?" and "How can we encourage a culture of innovation that embraces change rather than resists it?"

- **Ensuring security and ethical use of AI**: Taking a leaf from Google's book, prioritize the security of your data and the moral implications of AI deployments. This involves more than just compliance with regulations; it is about building trust with your customers and users. *Ask yourself,* "Have we implemented sufficient data protection measures to safeguard user privacy?" and "How do we ensure our AI systems are free from bias and respect ethical standards?"

- **Building robust systems for dependable decision-making**: Reflect on the robustness of PayPal's advanced AI algorithms and assess your systems' reliability, especially in critical applications. Implementing redundancies, error handling, and anomaly detection can enhance system reliability. *Ask yourself,* "How does our system handle unexpected inputs or conditions?" and "Can we maintain service quality under various stress scenarios?"

- **Achieving interoperability to enhance collaborative insights**: Adobe's Experience Platform illustrates the power of interoperability. Evaluate your ecosystem for its ability to integrate diverse data sources and systems. *Ask yourself,* "How seamlessly do our systems communicate and share data?" and "Can we easily integrate external data sources or third-party systems to enrich our insights?"

- **Maintaining an ecosystem to adapt and thrive**: Inspired by Nvidia, consider the long-term maintainability of your technology stack. Regular updates, modular designs, and precise documentation can simplify maintenance and ensure your ecosystem remains cutting-edge. *Ask yourself,* "How straightforward is it to update or enhance our systems?" and "Do we have a transparent process for staying abreast of technological advancements and integrating them into our ecosystem?"

The journey from strategic vision to practical implementation is complex and transformative. The expanded examples and these guidance questions should help develop more informed decisions. The path ahead is one of continuous learning, adaptation, and growth. Embrace it with the confidence that today's decisions will lay the foundations for tomorrow's successes and ensure that your organization remains resilient, agile, and aligned with the evolving landscape of data and AI.

Navigating the Future

The journey through data and AI architecture is more than a technical endeavor; it is a strategic exercise in aligning technological potential with visionary leadership. In this context, a leader's role bridges the gap between what technology can achieve and what their organizations need to thrive in a rapidly changing world. This requires a deep understanding of the possibilities presented by data and AI and their organizations' unique challenges and opportunities.

- **Strategic agility and the art of adaptation**: Agility is paramount. The landscape constantly evolves, with new technologies, methodologies, and ethical considerations emerging rapidly. Leaders must cultivate an environment of continuous learning and flexibility, encouraging teams to embrace change and experiment with new solutions. This does not mean chasing every technological trend but maintaining a strategic focus by using architectural principles to guide informed decisions to drive long-term value.

- **Ethics, responsibility, and the human element**: As data and AI become increasingly integral to business operations, leaders must navigate the ethical implications of these technologies. This includes considerations of privacy, bias, and the impact of AI decisions on individuals and society. Moral leadership in the age of data and AI means ensuring that technologies are deployed responsibly, with a clear commitment to positive outcomes for all stakeholders. It also means putting the human element at the center of technological initiatives, recognizing that behind every data point is a person, a community, or an ecosystem.

- **Fostering a culture of innovation and inclusion**: Creating a supportive ecosystem for data and AI extends beyond technology infrastructure; it involves cultivating a culture that values innovation, inclusion, and diversity. Leaders should foster environments where differing perspectives are welcomed and seen as essential to driving creativity and innovation. This includes breaking down silos between departments, encouraging cross-functional collaboration, and ensuring that all voices are heard and valued during the development process.

The path of digital transformation is ongoing, requiring leaders to remain vigilant, adaptable, and committed to their vision.

The Conscious Leader's Approach to Technology Ecosystems

In my journey—and that of countless other tech leaders I have encountered—implementing technology ecosystems has emerged as an operational endeavor and a profound journey of discovery and transformation. These experiences, rich with challenges and insights, reveal that the essence of technology transcends its mechanical functions, serving instead as a canvas for values, aspirations, and collective vision.

Take, for example, a leader of a burgeoning startup. Faced with the daunting task of architecting a technology ecosystem, they anchor their decisions on the core values of sustainability, community, and innovation. This conscious choice illuminates their path, transforming potential obstacles into opportunities for growth, collaboration, and a more profound connection with their team and the broader ecosystem they inhabit. By prioritizing sustainability, a leader ensures that their technology choices minimize environmental impact, using strategies such as opting for energy-efficient servers and promoting remote work to reduce carbon footprints. Community-driven decisions involve selecting open-source software that fosters collaboration and knowledge sharing. At the same time, focusing on innovation encourages the adoption of cutting-edge technologies that keep the company competitive.

An example of how these values translate into practical decision-making is evident in the startup's approach to cloud services. Rather than choosing the cheapest provider, a leader might opt for a cloud service that uses renewable energy sources to align with sustainability goals. They may also participate in community-driven projects or use platforms that support collaborative tools and frameworks to foster a sense of community within and outside the organization. Innovation is further reflected in their willingness to experiment with AI-driven analytics and ML tools to optimize operations and deliver superior customer experiences.

Such a narrative underscores a pivotal realization: that every choice, every investment, and every partnership in the realm of technology reflects leadership and the legacies created. By embedding core values into technology strategy, leaders build a resilient and forward-thinking ecosystem, inspire their team, and contribute positively to the broader tech community. This integrative approach ensures that technological advancements are in harmony with personal growth and collective aspirations, embodying the true essence of transformation.

Holistic Investment: Beyond the Balance Sheet

Beyond the cold calculus of investment returns lies a realm where the actual value of technology investments reveals itself. This is where the cultivation of an organization's culture, the enrichment of its human connections, and the capacity to inspire and engage happens.

Consider an organization that, in its quest for digital advancement, prioritized investments based on immediate technological merits and the potential to foster an environment of learning, innovation, and ethical engagement. This holistic investment approach revealed hidden value dimensions, transforming the workplace into a vibrant community of inspired individuals committed to a shared vision. The return on investment is measured not just in enhanced productivity or financial gain but in the flourishing of an organizational culture that champions creativity, integrity, and collective well-being. This expands how value is understood and recognizes that investments can shape the future of businesses, communities, and the world.

A Conscious Approach to Cost/Benefit Analysis

When creating a technology ecosystem, conscious leaders navigate the delicate interplay between visionary investments and tangible returns. This is not merely an exercise in fiscal prudence but a profound exploration of how each dollar spent mirrors deeper commitments to growth, sustainability, and ethical stewardship.

The process of assessing costs transcends the calculation of dollars and delves into the valuation of intangible assets—trust, brand equity, and the social impact of technological choices. It demands the consideration of the scalability and cost-effectiveness of infrastructure components and how they align with core values and collective well-being.

Similarly, evaluating benefits extends beyond immediate financial gains to enriching organizational culture, empowering teams, and nurturing innovation to serve humanity. This holistic view challenges us to redefine success in terms broader than mere competitive advantage and to consider how technology investments contribute to a legacy of positive change.

In navigating strategic decisions surrounding our technology ecosystem, the choice between building in-house solutions versus external technologies is a critical juncture. This decision is not merely a fiscal calculation but a reflection of alignment with strategic allies who share a common vision. Embracing the wisdom from *The Buddha and the Badass* leads to an alignment with the right partners and choosing external solutions that reflect broader values and goals, allowing an organization to concentrate on core strengths. This balanced approach fosters technological synergy and a shared journey toward a precise vision.

A Personal Reflection

Let me share a personal journey of mine. We were implementing an AI-powered chatbot for customer support within a venture I headed. The decision to do so aimed to increase operational efficiency and embody our ethos of customer-centricity and innovation. The initial costs—spanning software licensing, training, and integration—were weighed not just against the potential for reduced operational costs and enhanced productivity but also the opportunity to deepen our connection with customers through personalized and responsive service.

The ROI became a multidimensional measure, reflecting a financial metric and a gauge of customer satisfaction, team engagement, and the evolution of our brand, the data-collection firm DharmicData. Through its challenges and successes, the chatbot project became a testament to our journey toward integrating technology not as an end in and of itself but as a means to enrich our collective experience and journey toward growth.

Cost/Benefit Analysis

In the following hypothetical, the objective is to reduce customer service operational costs and improve response time to customer inquiries by incorporating an AI-powered customer support chatbot.

The following two tables lay out the scenario.

Category	Description	Initial Cost	Annual Recurring Cost
Research and Development	Vendor assessment, solution design, and initial training data collection.	$10,000	-
Software and Licensing	Chatbot platform licensing and additional software/tools.	$5,000	$5,000
Implementation and Integration	Integrating chatbot into the website, linking to database, setup.	$15,000	-
Training	Training staff to manage, supervise, and improve the chatbot.	$7,000	-
Maintenance	Monthly checks, software updates, and addressing any bugs/issues.	-	$12,000
Training Data Updates	Continually improving chatbot responses based on feedback.	-	$6,000
Total		$37,000	$18,000

Table. Cost Analysis for AI-Powered Customer Support Chatbot

Category	Description	Annual Benefit
Reduced Labor Costs	By utilizing the chatbot to handle 50% of customer inquiries, operational efficiency allows for reducing the customer service team or reallocating staff to other areas. This significantly reduces labor costs, with savings estimated from two full-time positions at $40,000 each.	$80,000
Increased Sales	Faster response times lead to a 5% increase in conversions from inquiries.	$50,000
Customer Retention	Improved response times and 24/7 availability lead to a 2% increase in customer retention.	$20,000
Operational Efficiency	Reduction in manual errors, standardization of responses, and freeing up human agents for complex inquiries.	(Qualitative)
Total		$150,000

Table. Benefit Analysis for AI-Powered Customer Support Chatbot

ROI Calculation

- **The ROI equation**:

ROI=(TotalBenefit−TotalCost)/TotalCost

- **For the first year**:

ROI = ($150,000 - ($37,000 + $18,000)) / ($37,000 + $18,000) = 0.4848

ROI = 0.48 or 48

- **For subsequent years (since initial costs are one-time)**:

ROI = ($150,000 - $18,000) / $18,000 =7.33733

ROI = 7.33 or 733

The AI-powered chatbot has a positive ROI, making it a worthwhile investment. The chatbot returns 48 percent of the invested amount in the first year. Given that the initial costs are not recurring, the ROI spikes to 733 percent in subsequent years. This quantified analysis and qualitative benefits like improved customer satisfaction make a strong case for green-lighting the project.

Quantifying ROI and Risk Mitigation

Organizations should conduct thorough cost/benefit analyses to make informed decisions about technology investments. Assigning monetary values to benefits and considering risk factors will help calculate the ROI and make a compelling business case for technology decisions.

However, ROI is only one metric for assessing technology decisions. Recognizing the value of intangible benefits such as customer satisfaction, employee morale, and strategic alignment with long-term organizational goals is essential. Moreover, risk analyses are critical in identifying potential setbacks, which enhances the proactive development of mitigation strategies.

As outlined in Chapter 8, KPIs encompass a 360-degree view of an organization's health and performance, with growth as a primary indicator. **Key examples include:**

- **Customer satisfaction**: Reflects how well the organization meets customer needs and expectations. High satisfaction rates often translate to customer loyalty and positive word-of-mouth.

- **Employee empowerment**: When employees are freed from repetitive tasks by automation, they can engage in more value-creating activities, leading to higher job satisfaction and innovative contributions.

- **Strategic alignment**: Initiatives must align with the organization's mission to ensure coherence and efficiency across all operations.

Integrating KPIs ensures that decisions regarding technology investments are holistic by considering a broad spectrum of impacts on the organization. Incorporating such considerations into cost/benefit analyses ensures that decision-makers have a clear view of the potential impact of technology investments. Such an informed approach leads to strategic decisions that underpin an organization's commitment to growth, innovation, and long-term success.

Technology decisions should not be made solely based on immediate needs. Organizations must consider their long-term impact to future-proof the technology ecosystem. Scalability, extensibility, and compatibility with emerging technologies are vital to ensuring the ecosystem's resilience and longevity.

A comprehensive cost/benefit analysis forms the bedrock of sound technology decision-making in the data and AI. By evaluating both the financial costs and the potential returns, organizations can strike the delicate balance between investing in innovative technologies and reaping the benefits of a thriving, value-driven technology ecosystem.

Fostering Deep Connections Through Technology Partnerships

Partnerships are more than mere transactional relationships when creating and nurturing a technology ecosystem. Finding the right partners involves identifying and collaborating with professionals who share your technical vision and align with your most profound mission and values.

A Conscious Approach to Selecting Technology Partners

When embarking on digital transformation, allies—our technology partners—play a crucial role. Drawing from the wisdom of the "Attract Your Allies" chapter in Lakhiani's *The Buddha and the Badass*, it is understood that partnerships are strategic and spiritual alignments that can amplify efforts, enrich the ecosystem as it is developed, and ensure that it serves immediate needs while also pursuing a grander vision. This approach to partnership invites looking beyond just technical capabilities and assessing the alignment of potential partners with a company's vision and ethical stance and the potential for coevolution that follows.

By integrating such a conscious approach, companies can build robust and symbiotic relationships with their technology partners, fostering a collaborative environment where shared values and mutual growth are prioritized. For instance, selecting partners who emphasize sustainability can reinforce a company's commitment to environmental responsibility, while partners who focus on innovation can help drive forward-thinking solutions. This alignment enhances operational efficiency and contributes to a more cohesive and purpose-driven organizational culture.

An Example of Conscious Decision-Making

For instance, in the story of a leader at the helm of a burgeoning startup, choosing a cloud services provider was not just about scalability and cost—it was about finding a partner committed to sustainability and innovation, a reflection of the startup's values. The leader chose a cloud provider that operated data centers powered by renewable energy, significantly reducing operations' carbon footprint. Additionally, the provider is firmly committed to research and development in green technologies, which aligns perfectly with the startup's innovation goals.

This shared vision laid the foundation for a partnership transcending the conventional. It created a collaborative force to drive societal and environmental change through technology. The startup and its cloud provider collaborated on projects to improve energy efficiency through AI-driven analytics and optimizing resource allocation. For example, they developed an ML model that dynamically adjusted server loads to minimize energy usage during peak and off-peak times. This lowered operational costs and contributed to broader environmental sustainability efforts. Moreover, they initiated a joint venture to create educational programs and workshops to promote sustainable technology practices within the industry, further amplifying their impact on environmental change.

By prioritizing partners with aligned values and a commitment to sustainability, the leader ensured that the technology ecosystem supported business growth and a positive societal and environmental impact. This approach to selecting technology partners exemplifies how strategic alliances can extend beyond business objectives to foster meaningful change worldwide.

Best Practices for Partnership Management

To manage technology partnerships with a holistic lens, **conscious leaders consider the following:**

- **Shared goals and values**: Ensuring that every collaboration serves the broader mission of positive impact, as emphasized in "Attract Your Allies." This means engaging in partnerships with entities that offer technological excellence and resonate with the organizational ethos of sustainability, innovation, and community building.

- **Collaboration beyond contracts**: Encourage a culture of shared learning and mutual growth. Technology partnerships should foster an environment where both parties can explore new territories, challenge each other, and grow together. This collaborative spirit mirrors the journey of finding and nurturing allies, as described in *The Buddha and the Badass*, creating a dynamic alliance capable of surmounting innovation challenges and market shifts.

- **Flexibility and adaptability**: Recognizing that achieving shared visions may require navigating unforeseen challenges together. In the spirit of "Attract Your Allies," maintain openness to evolving strategies and solutions in collaboration with partners, ensuring that mutual goals and values align.

Case Study: The Transformative Power of Strategic Partnerships

Let us consider a success story to illustrate the transformative power of strategic partnerships. Consider the anonymized alliance between Company X, a leading AI startup, and Company Y, a healthcare provider.

Company X (known for its innovative AI solutions) and Company Y (a healthcare provider focused on patient-centric care) formed a strategic partnership to address critical healthcare challenges using AI. Their collaboration aimed to develop a platform to enhance patient diagnostics and personalize treatment plans. This project involved integrating Company X's advanced ML algorithms with Company Y's extensive patient data to create predictive models that could identify potential health issues before they become critical.

Company X provided technological expertise and innovative solutions through the partnership, while Company Y offered invaluable industry knowledge and patient data. This synergy drove the creation of a platform that improved diagnostic accuracy and optimized treatment plans, leading to better patient outcomes and the more efficient use of healthcare resources.

This success story demonstrates how strategic partnerships can leverage both parties' strengths to create more significant solutions than the sum of their parts. By aligning their goals and working together, Company

X and Company Y were able to drive significant advancements in healthcare technology, illustrating the profound impact that such collaborations can have.

Case Study: Conscious Selection and Management

Let us consider another anonymized case study that illustrates the conscious selection and management of technology partnerships.

EcoFuture Solutions, a startup focusing on sustainable technology solutions, sought a cloud computing provider to host their innovative environmental analytics platform. They chose CloudGreen Innovations, which is known for its commitment to renewable energy and eco-friendly data centers.

- **Alignment of vision and values**: Both companies shared a deep commitment to environmental sustainability, making the partnership a business decision and a mutual commitment to positive global impact.

- **Collaborative growth and learning**: EcoFuture and CloudGreen engaged in knowledge exchange programs, where experts shared insights on sustainable computing practices, enhancing the understanding and capabilities of each.

- **Adaptability in partnership**: As EcoFuture's needs evolved, CloudGreen adjusted services and resources flexibly, demonstrating the partnership's dynamic nature and mutual commitment to supporting growth and innovation.

This case study exemplifies the essence of "Attract Your Allies" in practice, highlighting how strategic partnerships, grounded in shared values and visions, can amplify the transformative impact of technology initiatives.

Reimagining how to manage technology partnerships involves not just leveraging external capabilities but embarking on a journey of collective transformation. Grounded in shared values and visions, this process can usher in a new era of technological innovation that honors the essence of human connection and the spirit of societal advancement.

Inspired by *The Buddha and the Badass*, follow a path that includes actively engaging in industry consortia, contributing to open-source projects, and forming strategic alliances. This communal approach enriches the potential to innovate and establishes the reputations of thought leaders within this vibrant ecosystem. Such collaborative efforts amplify impact, and technological advancements are woven into a collective vision for the future.

Anticipating the Next Steps of Data and AI

Beyond the confines of conventional wisdom lies a realm where emerging technologies converge to redefine digital existence's very fabric. The future knows no bounds, from the transformative potential of blockchain and decentralized technologies to the paradigm-shifting capabilities of quantum computing.

However, amidst the allure of novelty, discernment must be practiced. Not all that glitters is gold. Each technological leap comes with its own set of challenges and ethical considerations.

Quantum Computing

Quantum computing, often heralded as the next technological frontier, diverges fundamentally from classical computing that uses bits (0s and 1s) to process information. Quantum computers employ quantum bits or qubits that can represent 0 and 1 simultaneously. This principle, known as superposition, allows quantum computers to perform many calculations at once.

Quantum computing holds the potential to revolutionize various industries. According to McKinsey & Company, its potential market is expected to reach $1 trillion by 2035. This showcases its expected adoption and massive influence across various sectors (McKinsey & Company n.d.).

However, the concept of entanglement is another quantum principle that truly sets quantum computing apart. In this state, the information of one qubit depends on another—even if they are light-years apart. This interconnectedness can vastly increase computational power. **The implications of quantum computing are manifold and profound:**

- **Cryptography**: Most of today's online security protocols are built on the difficulty of factoring large numbers into primes, a task manageable for classical computers—though time-consuming. Quantum computers could crack these encryptions in seconds, prompting a complete overhaul of cybersecurity measures.

- **Drug discovery**: Quantum computers can rapidly simulate complex molecular and chemical reactions, accelerating the discovery of new drugs and materials.

- **Optimization problems**: From traffic flow optimization to perfecting supply chain logistics, quantum computers could solve problems deemed too complex for classical computers.

Federated Learning

In traditional machine learning, data is centralized in a single location for model training. This poses potential risks related to data privacy and transfer costs. Federated learning turns this paradigm on its head. Instead of bringing data to the model, it brings the model to the data. In this decentralized approach, individual devices (like smartphones or devices connected to the IoT) train ML models locally using their data. These local models then share updates or improvements with a central server, aggregating this knowledge and updating the global model.

The worldwide federated learning market is projected to reach $311.4 million by 2032 (Market.us 2024), driven by factors such as the increasing adoption of federated learning in various industries and the rising demand for secure and efficient data-sharing practices. This growth reflects the recognition of its potential to address privacy concerns while enabling collaborative model training across different organizations.

A study published in *Engineering Proceedings* emphasizes the importance of federated learning in enhancing data privacy and security while allowing collaborative model training across different organizations (Dhade & Shirke, 2023). The research highlights how it can revolutionize industries like healthcare by allowing multiple parties to train models without sharing sensitive data, leading to advancements in precision medicine and personalized treatments. These insights underscore the significant potential of federated learning to transform how organizations approach machine learning and data privacy in the coming years.

The reach and influence of federated learning span various domains:

- **Data privacy**: Since raw data remains on the local device and only model updates are communicated over the Internet, user data is less prone to breaches, ensuring robust privacy safeguards.

- **Bandwidth efficiency**: Eliminates the need to continuously send vast amounts of data to centralized servers, conserving bandwidth and reducing data transfer costs.

- **Real-world applications**: From enhancing predictive texting on smartphones without sending data to the cloud to enabling wearable health devices to monitor and predict health anomalies better, federated learning is poised to revolutionize how AI integrates into our daily lives.

Other Emerging Technologies

Several other possibilities are also coming into view:

- **Explainable AI (XAI)**: The demand for transparency in AI is increasing as explainable AI becomes more crucial in critical decision-making processes. Developing techniques that provide human-understandable explanations for AI decision-making is essential for addressing AI bias and complying with regulatory demands.

- **Machine learning bias**: A deep understanding of data and its quality is necessary to prevent the inherent biases in ML models. The call for a more profound knowledge regarding data to avoid bias has become more pronounced.

- **Cloud and edge computing**: The synergy between cloud and edge computing is growing, with a deeper understanding of how they can work together to enhance real-time data processing and decision-making. Edge computing refers to processing data closer to where it is generated instead of sending it to centralized data centers. This approach reduces latency and bandwidth usage, enabling faster and more efficient data processing, which is crucial for applications requiring real-time insights.

- **Bioengineering**: Technologies ranging from clustered regularly interspaced short palindromic repeats (CRISPR) and genomics to cultured meat production are increasingly being explored. Bioengineering is gaining traction, especially in health, vaccines, consumer goods, and materials.

As the vast landscape of emerging technologies is further explored, it is clear that integrating these innovations into data and AI strategies is essential for staying competitive and innovative. However, successfully adopting these technologies is not just about understanding their potential but also about implementing them wisely and responsibly.

Best Practices: Navigating the Journey

Crafting a resilient technology ecosystem is essential for organizations across all industries, yet applying these best practices may vary depending on specific sector requirements. While the principles outlined below are broadly applicable, their implementation can be tailored to meet the unique challenges and opportunities of distinctive fields, such as healthcare, finance, retail, and manufacturing. Deliberate planning and informed decision-making are universal keys to success.

- **Holistic technology evaluation**: Evaluate technologies in isolation and as integral parts of the larger ecosystem. This ensures seamless integration and compatibility between components in a highly regulated industry like finance, healthcare, or a fast-paced retail environment.

- **Agility and scalability**: Technologies that allow seamless scaling and rapid adaptation to changing business demands are crucial across sectors. Avoiding rigid solutions is especially important in industries like technology and retail, where market demands can shift quickly.

- **Security by design**: Integrating security measures at every layer of the technology ecosystem is vital in all fields. However, it is particularly critical in sectors dealing with sensitive information, such as healthcare and finance.

- **Automation and monitoring**: The drive towards automation and robust monitoring is universal, though the specific applications and technologies might differ by industry. In manufacturing, for instance, this might involve real-time monitoring of production lines.

- **Talent and skill development**: Investing in team development is a best practice across industries. Continuous learning and experimentation keep your organization at the forefront of technological advancements, regardless of your sector.

User-Centered Design (UCD)

Prioritizing the end user's experience is a universal best practice that is particularly relevant across industries. According to Forrester, a well-designed user interface could raise your website's conversion rate by up to 200 percent, and a better UX design could yield conversion rates up to 400 percent. This data underscores the value of UCD across sectors, from e-commerce platforms seeking to boost sales to healthcare applications that enhance patient engagement.

Prioritizing Data Privacy and Security

With the increasing prevalence of data breaches and the tightening of data protection regulations worldwide, prioritizing data privacy and security is necessary for all industries. As highlighted in various industry reports, including IBM's 2021 *Cost of a Data Breach Report*, the impact of an average data breach reached a record level of $4.24 million in 2021, which emphasizes the critical nature of data security across all sectors. This statistic is a stark reminder of the importance of adopting rigorous data protection practices from the outset, irrespective of industry.

Integrating these best practices—tailored to an industry's unique demands and enriched with a clear understanding of their importance and credibility—prepares companies to navigate and thrive within the digital era (Anderson 2021; Bluefin 2023; IBM Security n.d.; Sobers 2023; Jerzewski 2023).

Conclusion

The future is digital, interconnected, and fueled by data. However, working towards this demands more than passive observation—it calls for deliberate, proactive action.

Assembling a technology ecosystem for data and AI is a monumental undertaking, but organizations can succeed by adhering to architectural principles, leveraging appropriate technologies, and incorporating best practices. The real-world case studies explored in this chapter exemplify the significance of these decisions, demonstrating how resilient technology ecosystems drive tangible business outcomes.

An organization's future lies in decisions made today:

- **Assess and reflect**: Begin by assessing the current technological ecosystem. Where do things currently stand, and what are the goals moving forward?

- **Engage and collaborate**: Bring together the best minds—technologists, strategists, and business leaders. Foster a culture of collaboration where diverse perspectives shape the blueprint for the tech future.

- **Educate and adapt**: The technological landscape evolves rapidly. Invest in continuous learning and be ready to adapt to shifts, whether those are emerging technologies or industry disruptions.

- **Implement with a vision**: Do not just chase the newest shiny tech trend—ground technology decisions in a well-articulated vision that aligns with an organization's mission and long-term goals.

- **Iterate and evolve**: Perfection is a journey, not a destination. An ecosystem will need tweaks, adjustments, and, sometimes, significant shifts. Embrace this continuous evolution.

In the digital age, standing still is moving backward. The principles of modern data and AI architecture and narratives of strategic partnerships underscore a profound truth: constructing a technology ecosystem is both a scientific endeavor and a philosophical quest that reflects our deepest commitments to innovation, ethical stewardship, and meaningful societal impact.

The alliances formed while moving forward, rooted in shared values and mutual development, emphasize that the essence of our technological journey is intrinsically human. These partnerships extend beyond the boundaries of conventional transactions, evolving into collaborative endeavors toward strategic objectives that also nurture a culture of innovation and inclusion.

CHAPTER 10

Establishing a Solid Base Camp

This chapter covers the essential preparations for a successful ascent in the digital age—establishing the base camp. That means the critical components of data management, from ensuring the integrity and accessibility of data to safeguarding its quality and meeting compliance requirements. The complex terrain of data architecture will also be identified, from centralized and distributed systems to innovative concepts like data mesh and lakehouses. The expedition is about more than just technical mastery; it is about the human element of data management.

Data Management Components: People, Processes, and Workflows

Data management components are the bedrock of reliable and trustworthy data. As AI capabilities advance, components like data governance, master data, and metadata are primed for continuing transformation. AI promises to elevate and reimagine entire data management components by automating repetitive tasks and applying intelligence to data oversight.

Master Data Management (MDM)

This is the cornerstone of an organization's data strategy, essential to overcoming the challenges posed by data silos—separate islands of data within an organization—that create barriers to operational efficiency and coherent decision-making. By centralizing core data entities like customers, products, and suppliers into a single, authoritative source of truth, MDM addresses the inefficiencies and limitations imposed by data silos.

The process of dismantling barriers to data flow involves several key steps:

- **Integration**: MDM integrates data from various sources, addressing the inconsistencies and duplications that often occur when multiple systems collect and store the same data independently.

- **Cleansing**: Once integrated, data is cleansed to correct inaccuracies, remove duplicates, and fill in missing values to ensure all stakeholders can access high-quality data.

- **Standardization**: MDM standardizes data formats, definitions, and terminologies across the organization, facilitating clear communication and uniform understanding among all departments.

- **Governance**: MDM establishes a governance framework that defines who can access data, how it can be used, and who is responsible for maintenance. This framework ensures that data is used appropriately and remains accurate over time.

By executing these steps, MDM creates a streamlined operational environment where decision-makers can access reliable data. This reliability provides a clear, comprehensive, and current view of critical business entities. For instance, a global retail company (an anonymized former client) experienced significant improvements in business outcomes by unifying customer data from disparate systems. This unification allowed the company to identify and understand customer behaviors and preferences more clearly, enabling tailored marketing campaigns that led to a 20 percent increase in customer retention and a 15 percent improvement in cross-selling revenue.

In practice, the MDM's unification of customer data from disparate systems meant that the retail company could identify and understand its customers' behaviors and preferences more clearly. This deepened understanding allowed the company to tailor marketing campaigns more effectively, leading to a more personalized customer experience. Personalization is a well-documented way to enhance customer satisfaction and loyalty, hence the observed 20 percent increase in retention. Additionally, with a more comprehensive view of each customer, opportunities for relevant cross-selling became more apparent, resulting in a 15 percent improvement in cross-selling revenue.

When implementing MDM, organizations create a centralized, interconnected data ecosystem that aligns with their core mission and values. Viewing MDM through this lens involves technical implementation and building a data structure that reflects and supports an organization's highest purpose and aspirations.

Case Study: Office Depot

Office Depot, a leading global provider of office products and services, recognized the need to unify customer data across multiple channels and geographies to improve business performance. The company's vast and varied product information was scattered across different systems, leading to inconsistencies and a fragmented customer experience.

In response, Office Depot implemented an MDM solution to consolidate and manage its extensive product data. This initiative was part of a broader omnichannel strategy to ensure consistency across all customer touchpoints (Talend, n.d.; Karatas, 2024; Shopsys, n.d.; Cox, 2001; Forrester, n.d.).

Key outcomes from Office Depot's MDM implementation included:

- **Unified product information**: Office Depot established a single source of truth for product data, which resulted in more consistent and accurate product information across all sales channels.

- **Enhanced customer experience**: By providing accurate and detailed product information, Office Depot could offer customers better search functionality, more precise product descriptions, and reliable availability data, leading to a more informed and satisfying shopping experience.

- **Operational efficiency**: MDM streamlined internal processes, reducing the time and resources required to manage product data. This efficiency gain allowed Office Depot to reallocate resources to more value-adding activities.

- **Improved sales**: As a direct result of implementing MDM, Office Depot experienced an improvement in customer retention due to the enhanced shopping experience. They also saw increased cross-selling and upselling opportunities, as accurate product data enabled more effective marketing and sales strategies (Shopsys, n.d.; Forrester, n.d.; Tacken, 2019; Venkataraman & Cross, 2009; Cox, 2001).

Metadata Management

This is the foundation upon which data-driven organizations build their ability to utilize vast amounts of data efficiently. By cataloging the details and characteristics of data, metadata management facilitates easier access, understanding, and governance of data assets. Enhancing data quality, supporting data lineage, and ensuring standardization and interoperability across an organization is critical.

Case Study: Netflix

Netflix, a global leader in streaming entertainment, has mastered metadata management and enhanced its content delivery and personalization algorithms. With a vast library of movies, TV shows, and original content, Netflix employs a sophisticated metadata tagging system to categorize and recommend content to users based on their viewing preferences (Turon 2023; Fernández-Manzano, Neira, & Clares-Gavilán 2016; Atlan 2023).

Key metadata management achievements at Netflix include:

- **Improved content discovery**: The metadata system tags content with various descriptors, from genre and actors to mood and themes. This granular detail provides highly accurate recommendations, making content discovery a seamless user experience.

- **Enhanced personalization**: The rich metadata allows Netflix to tailor its user interface and content recommendations to individual preferences, significantly improving user engagement and satisfaction.

- **Operational efficiency**: Metadata management streamlines the process of adding new content to the platform, ensuring that all necessary information is captured and accurately represented. This efficiency reduces the time to market for new content and enhances the overall content management process.

- **Data governance and quality**: Netflix's metadata framework supports robust data governance practices, ensuring that metadata is consistently applied and maintained across its content library. This consistency is crucial for the integrity of content discovery and recommendation algorithms.

By integrating comprehensive metadata management into its operations, Netflix has optimized its content delivery and personalization and set a benchmark in the entertainment industry for leveraging data to enhance user experience.

Data Governance

This is the linchpin in an organization's data strategy, a force maintaining order and ensuring that data serves the organization's collective goals responsibly and effectively. A set of standards and practices provides the framework for data management, encompassing quality, privacy, and accessibility. Good data governance aligns data management with the organization's strategic objectives, ensuring that data is an asset and a beacon guiding decision-making processes.

The fundamental impacts of sound data governance include:

- **Assured data quality**: Data governance ensures that data is accurate, complete, and reliable via rigorous standards and procedures.

- **Enhanced data security**: Robust governance protocols safeguard sensitive data against unauthorized access and potential breaches.

- **Ensured compliance:** Data governance is essential for meeting regulatory requirements and helping organizations navigate the complexities of compliance with laws like GDPR, HIPAA, and others.

- **Effective risk management**: Proactive governance strategies mitigate the risks associated with data management, particularly the incorporation of third-party data.

Data governance can be seen as a way to establish the principles and beliefs defining an organization's interaction with data. Just as spiritual practices often involve guidelines or ethics to live by, data governance ensures that data is used ethically and serves the highest good for all stakeholders. This approach elevates it from rules to reflect the organization's values and commitment to integrity.

Case Study: Sutter Health

Sutter Health is a not-for-profit health system based in Northern California that faced navigating the complexities of data governance across its extensive network of facilities and systems. Its journey underscores the critical need for robust data governance to maintain high data quality and regulatory compliance amidst the healthcare industry's evolving demands and challenges (Sutter Health n.d.; California v. Sutter Health System 2001; Gudiksen 2019; US Department of Justice 2021; Powderly 2015; Macht 2023; Downs 2023; Advisory Board n.d.; R1 2022). **Data governance efforts at Sutter Health include:**

- **Unified data standards**: As part of its data governance journey, Sutter Health has been working towards establishing common data definitions. These efforts aim to enable more consistent and reliable healthcare data, though challenges such as data breaches have highlighted areas for improvement in security practices.

- **Data security initiatives**: In response to vulnerabilities, Sutter Health has been focused on enhancing data access controls and security measures. In particular, it experienced a significant data breach in 2011, exposing the personal information of over 4 million patients. This incident underscored the urgent need to tighten security protocols and access controls. While these initiatives are steps toward safeguarding patient information, they also reflect ongoing efforts to strengthen data protection in light of historical challenges.

- **Striving for regulatory adherence**: Sutter Health's data practices have been developed to align with industry standards and regulations. It has faced multiple legal challenges, including a $575 million settlement with the State of California over allegations of anti-competitive practices and compliance issues with healthcare laws. The health system has faced legal scrutiny and settlements related to healthcare laws, underscoring the importance of continuous improvement and adherence to regulatory requirements.

- **Addressing operational risks**: One operational risk they encountered was the need for standardized data management policies, leading to inefficient data usage and decision-making processes across different departments. These measures contribute to the organization's journey towards improved operational integrity despite facing operational risks and the need for enhanced data governance frameworks.

This case study of Sutter Health illustrates the complexities of implementing effective data governance within a large healthcare organization. While Sutter Health has initiated steps to manage data quality, security, and compliance, the organization's experiences with data breaches and legal challenges are essential reminders of the need for vigilance, continuous improvement, and a culture of accountability in data governance practices, emphasizing the need to establish a solid framework for managing data as an ongoing process that can lead to organizational learning and improvement.

Data Quality

The heartbeat of a data-driven organization—data quality—ensures that the lifeblood of decision-making is pure and reliable, meaning it is accurate, complete, consistent, and dependable. That it can act as a trustworthy foundation for analytics and operations. This encompasses a series of processes and governance practices to measure, monitor, and manage the quality of the data flowing through an organization while fostering a culture of ownership and quality consciousness among data professionals.

The key impacts of prioritizing data quality include:

- **Trustworthy analytics**: High-quality data underpins credible analytics, leading to sound business insights and decisions.

- **Operational efficiency**: Accurate and consistent data streamlines operational processes, eliminating errors and redundancies that can cost time and resources.

- **Customer satisfaction**: Reliable product data ensures customer expectations are met, which enhances the customer experience and brand loyalty.

- **Regulatory compliance**: Maintaining data quality is often a legal requirement, particularly for financial reporting and personal data handling.

Case Study: Best Buy

Best Buy, A leading electronics retailer, has consistently sought ways to enhance its e-commerce platform and overall customer experience. Although specific details about a particular data quality initiative at Best Buy are not publicly documented, the focus on maintaining accurate product information across all channels is a common goal among large retailers like Best Buy (Campbell 2024; 3Pillar Global 2021**). Some approaches to improving customer experience are:**

- **Accuracy in product information**: Any retail giant must ensure accurate product descriptions and specifications across various channels. Like other major retailers, Best Buy invests in systems to standardize product data, ensuring consistency and reliability. This accuracy is vital for customer trust and satisfaction, leading to better-informed purchase decisions.

- **Enhanced customer experience**: Initiatives to improve the customer experience, whether through data quality improvements or other methods, are essential. Best Buy implemented various enhancements— such as improved website navigation, personalized recommendations, and responsive customer service—contributing to a seamless shopping experience. The retail industry widely recognizes the link between accurate product information and enhanced customer experience.

- **Customer loyalty and revenue growth**: For retailers, initiatives that improve the customer experience indirectly influence loyalty and revenue. Accurate and reliable product information can reduce customer service inquiries, leading to higher satisfaction and potentially driving repeat purchases. Best Buy's efforts to ensure data accuracy and enhance the customer experience reflect broader industry practices that contribute to sustained growth and customer retention.

Accurate product information—enhanced through quality management systems—can lead to a better customer experience, increased loyalty, and, potentially, revenue growth. Retailers like Best Buy continuously explore ways to leverage data quality for competitive advantage, emphasizing the broader relevance of these initiatives in the digital transformation journey.

Data Catalogs

Imagine a vast celestial library where every book is a dataset, and every page details specific information about the universe. Data catalogs play a similar role in data management, acting as centralized repositories that organize, categorize, and make data discoverable across an organization. By indexing data assets with metadata, data catalogs not only make data easily accessible but also understandable, ensuring users can find and interpret the information they need. **A well-implemented data catalog facilitates:**

- **Enhanced data literacy**: This democratizes data access, enabling all users, regardless of their technical expertise, to understand and leverage data in their roles.

- **Promotion of collaboration**: By making data discoverable and transparent, data catalogs break down silos, fostering collaboration and innovation across departments.

- **Increased efficiency**: The time and effort spent searching for data are reduced, streamlining the analytics and decision-making processes.

- **Empowerment of self-service analytics**: Data catalogs empower users to access and analyze data independently, reducing the burden on IT departments and accelerating insight generation.

Case Study: Airbnb

Amid rapid expansion, Airbnb encountered challenges related to data management, scalability, and ensuring data consistency across its global platform. Addressing these issues was crucial to supporting the company's growth and maintaining high-quality service for hosts and guests.

To tackle these challenges, the internal tool Dataportal was developed. This innovative platform was designed to democratize data access within the company, allowing employees from various departments to discover, acquire, and utilize data efficiently. By facilitating smoother data interaction, Dataportal played a crucial role in transforming Airbnb's approach to data management and significantly contributed to a culture of informed decision-making (Saradhi, 2023; Zeenea, 2022; Al Farabi, 2023a). **Key outcomes from the implementation of Dataportal included:**

- **Centralized data discovery**: Provided a centralized hub for data discovery, enabling employees to find needed data quickly. This reduced search times and eliminated data silos, making data more accessible across the organization.

- **Enhanced productivity**: Allowed teams to work more efficiently by simplifying access to data and insights. Productivity improvements resulted from faster project turnaround times and more agile responses to market demands.

- **Improved data quality**: Provided better oversight and governance, enabling Airbnb to enhance its data quality. This led to more accurate listings on the platform, fewer customer service inquiries, and fewer booking cancellations due to data errors.

- **Data-driven culture**: Facilitated a shift towards a more data-driven culture at Airbnb. Employees were empowered to leverage data for decision-making, leading to more informed strategies and actions that supported business growth and efficiency.

Airbnb's development and implementation of Dataportal highlight the importance of effective data management to support scalability and efficiency at tech-driven companies. By addressing data challenges head-on, Airbnb improved its operational efficiency and strengthened its position as a leader in the online marketplace for lodging and tourism experiences.

Data Lineage

Resembling a cosmic map where every celestial body's position and path are charted, data lineage represents the journey of data through its lifecycle—its origins, transformations, and destinations. Such intricate mapping is crucial for understanding how data moves and changes across the vast expanse of an organization's operations, ensuring data integrity, accuracy, and compliance with regulations. Successful data lineage systems require robust collaboration and communication, uniting different teams around a unified vision of the organization's data flow. **Key benefits of establishing clear data lineage include:**

- **Improved data quality**: By tracing data back to its sources, organizations can quickly identify and rectify errors, enhancing the overall quality of the data.

- **Enhanced regulatory compliance**: Data lineage provides the kind of clear trail that regulatory bodies often require for compliance, simplifying audits and reporting.

- **Efficient troubleshooting**: Understanding data flow makes it easier to pinpoint where issues occur, significantly reducing resolution times.

- **Informed decision-making**: With a comprehensive view of the data's journey, decision-makers can rely on the accuracy and relevance of the data they use, leading to more informed business decisions.

Case Study: eBay's Data Lineage System

Managing an intricate and expansive data ecosystem with advanced IT tools, business analytics, and big data technologies like Apache Hadoop, eBay deployed a sophisticated data lineage system to significantly enhance its capacity to process vast quantities of user-generated data efficiently. The system is instrumental in providing comprehensive details about each data element to facilitate informed decision-making processes within the company. **Key achievements of eBay's data lineage system include:**

- **Data tracking at scale**: Designed to meticulously track the movement and transformation of data across its platform, the system ensures transparency and accuracy in understanding data provenance and flow, which is critical for a marketplace of eBay's magnitude.

- **Rapid resolution of data quality issues**: The data lineage system's precise mapping capabilities allow eBay to quickly identify and resolve data quality issues. This fast response capability minimizes potential impacts on business operations and customer experiences.

- **Regulatory compliance**: eBay operates across various jurisdictions, each with its data protection regulations. The data lineage system supports adherence to these regulations by providing a clear, auditable trail of data movements and usages.

- **Enhanced data-driven decision-making**: The trust established in data accuracy and completeness by the data lineage system empowers eBay's analysts and decision-makers. This trust facilitates strategic initiatives based on reliable data, driving the company's continuous growth and innovation.

eBay implementing a data lineage system exemplifies the transformative impact that a comprehensive understanding of data's journey can have on an organization's ability to manage data quality, ensure regulatory compliance, efficiently troubleshoot issues and make informed decisions based on reliable data.

Data Management Discipline	Key Takeaways	Example Impact
Master Data Management	Create a single, consistent, authoritative source of truth for critical data entities. Eliminate data silos to streamline operations and enhance decision-making.	A retail company implemented MDM to merge customer data, leading to improved marketing strategies and customer relations.
Metadata Management	Manage essential information about data to ensure data lineage and facilitate data discovery. Enable efficient data governance and data understanding across the organization.	A financial services firm introduced metadata management, resulting in more efficient data handling and quicker insight generation.
Data Governance	Establish policies, processes, and roles to ensure data quality, security, and compliance. Manage data as a valuable asset to foster trust in data-driven decision-making.	After enforcing data governance policies, a healthcare organization significantly reduced data breach incidents and aligned with compliance requirements.
Data Quality	Assess and improve data accuracy, completeness, consistency, and reliability. Ensure reliable analytics and meaningful insights.	An online retailer's investment in data quality led to a noticeable decrease in customer complaints and a boost in overall satisfaction.
Data Catalogs	Create centralized repositories for data assets to enhance data accessibility and	Introducing a data catalog in a tech company catalyzed self-service data exploration, markedly enhancing cross-departmental collaboration.

Table. Key Takeaways for Data Management Components

Mastering Data

Data is the compass that guides decisions, strategies, and innovations. Mastering data management components—such as metadata management, data governance, and data quality—ensures that the compass remains accurate and reliable. This empowers charting a course through uncharted territories, discovering new paths, and adapting strategies to the ever-changing conditions.

166

Human Elements in Data Management

The spirit of mountaineering is recognizing that every climber's strengths and contributions are vital to the expedition's success. Likewise, in data management, the human element is paramount. The journey depends not solely on the tools and technologies employed but on the people who wield them. As part of establishing a base camp, it must be recognized that skilled data professionals, effective workflows, and collaborative efforts are as crucial as the technical infrastructure. These elements ensure that data management practices are efficient and adaptive to the evolving landscape that must be overcome during the climb.

At its core, the journey towards effective data management and the use of AI is deeply human, built on understanding, creativity, and collaboration. Fostering a data-centric culture requires acknowledging the human element—imagination, intuition, and capacity for growth. It is about leveraging technology for economic gain and as a tool for enhancing the human experience and contributing to the collective well-being.

Ascending the Data Summit

Like preparing for a mountain climbing expedition, establishing a robust data foundation requires foresight, planning, and a strategic approach that will be implemented in phases and adapt to the terrain to ensure a successful ascent. **Here is a practical guide—incorporating the theoretical underpinnings explored so far—of actionable steps in an organization's data and AI journey:**

- **Phase 1—Laying the groundwork**: Just as climbers select a mountain and route, organizations begin their journey focusing on high-impact data domains—customer, product, and financial data—that are the bedrock for crucial business processes. This phase involves identifying and prioritizing these domains and establishing master data management, data quality, and governance frameworks to develop minimum viable capabilities (MVC). Think of MVCs as the climbing gear necessary for the initial part of the ascent.

- **Phase 2—First ascent**: Now, foundational capabilities are aligned with quick-win use cases to provide immediate value. In this phase, like finding the best path through the foothills, actionable insights are identified through collaborative exploration with business teams. These insights are refined through iterative development, starting with a proof of concept and building toward a minimum viable product (MVP) for pilot testing.

- **Phase 3—Establishing base camps**: In the scaling phase, efforts are expanded by building upon the customer data domain and gradually including other domains, such as financial and employee data. This is like establishing successive base camps further up the mountain, each serving as a launch pad for exploring new paths and enabling the organization to broaden its data management capabilities.

- **Phase 4—Acclimatization**: With the ascent underway, attention turns to acclimatizing the organization to a data-centric culture. This phase involves training, workshops, and communication initiatives to ensure every team member understands the importance of data and is prepared to climb. Like climbers adjusting to the altitude, organizations need time to embed new practices into their operations.

- **Phase 5—Advanced tools and navigation**: Just as climbers need more advanced equipment for the steeper parts of the mountain, organizations now integrate sophisticated tools like data catalogs and lineage documentation. These tools offer the means to navigate the complexities of data management, automating and streamlining the process, much like GPS and modern climbing gear assisting a safe and efficient climb.

- **Phase 6—Summit push**: As the journey nears its culmination, the maturity of data management capabilities is assessed to identify improvement areas and ensure that data practices yield tangible business outcomes. This is akin to the final push to the summit, where climbers must evaluate their progress, adapt strategies, and overcome the last hurdles to reach the peak.

Remember that the expedition metaphor is about the process as much as the destination. The organization's growth and the insights gained from every stage of the journey enrich its capabilities and competitive edge.

While all data management components are necessary, their implementation is not one-size-fits-all. They need to be tailored to an organization's needs and maturity level. This is about choosing the right gear for the climb, ensuring that each piece serves a purpose, and adapting to the conditions of the mountain. The art lies in balancing foundational work with progressing toward use cases, creating a rhythm that advances the organization's data capabilities while delivering value at each step.

Data architecture is a landscape that shifts with the winds of business needs and technological innovation. Recognizing that established principles are our foundation, we apply them with renewed creativity to meet today's challenges.

Data Architecture Trends

Data architecture serves as the backbone of modern enterprises, guiding the flow of information and unlocking the potential of data-driven decision-making. Far from being a static blueprint, the art of data architecture adapts and grows, evolving with the shifting terrains of business requirements and technological advancements. Established best practices provide a reliable map, but the spirit of innovation charts the path to new horizons.

Applying Established Concepts to Contemporary Challenges

The essence of data architecture is translating sophisticated methodologies into clear, actionable strategies. What follows is a focus not on unproven novelties but on the practical application of time-honored concepts tailored to the challenges of the digital era.

- **Centralized vs. distributed data**: The interplay between centralized governance and the flexibility of distributed systems is a way to seek a balance tailored to an organization's objectives and challenges.

- **Data mesh**: This embraces a shift towards domain-driven design and pragmatic steps for implementing a decentralized data ownership model to enhance the autonomy and agility of domain-specific teams without necessitating a wholesale architectural overhaul.

- **Lakehouse**: A unified analytics framework that combines the best of both worlds can be established by combining the vast storage capabilities of data lakes with the structured querying power of data warehouses.

- **Data fabric**: Disconnected data streams can be woven into a seamless quilt of information, enabling swift and strategic decisions across the enterprise.

Various data architectures can be likened to different spiritual paths or practices. Each offers a unique way to manage and leverage data, just as different spiritual teachings offer various paths to growth and understanding. The choice of architecture depends on the organization's specific needs and goals, similar to how an individual selects a spiritual practice that resonates with their journey.

The architectural paradigms discussed are more than theoretical constructs; they are practical insights forged in the furnace of real-world applications. Each model comes with its own set of trade-offs and decisions. These complex choices must be navigated to craft an architectural strategy that aligns with an organization's unique direction and aspirations.

Centralized vs. Distributed Data

Diverse organizational needs demand different architectural approaches. While centralized systems simplify management and unify data sources, they may need help scaling with increasing data volumes. In contrast, distributed systems, like data lakes, offer expansive storage and processing capabilities while demanding a rigorous governance framework to ensure data integrity.

In a centralized architecture, all data is stored in a single, unified repository, such as a data warehouse. This approach simplifies data management, ensures data consistency, and facilitates more straightforward analysis. However, it may face scalability challenges as the volume of data grows.

On the other hand, a distributed architecture like a data lake allows organizations to store data in its raw, unstructured form, accommodating large volumes of data. This setup supports diverse data types and enables parallel processing, but it requires a robust data governance framework to maintain data quality and security.

Parameter	Centralized Architecture	Distributed Architecture
Definition	A single, centralized system handles all data and operations.	Multiple interconnected systems share and handle data and operations.
Advantages	- Simplicity in management - Single source of truth - Easier data governance	- Scalability - Flexibility - Redundancy - Can better handle large loads
Disadvantages	- Scalability issues - Single point of failure - Can be a performance bottleneck	- Complexity in management - Data consistency challenges - Higher initial setup cost
Best For	- Smaller organizations - Simple workflows - Less diverse data sources	- Large organizations - Complex, data-intensive applications - When high availability is required
Real-World Example	Traditional relational database systems like Oracle DB, Microsoft SQL Server	Systems like Apache Kafka, Apache Cassandra, and microservices-based applications like Netflix

Table. Comparison of centralized vs. distributed architectures

Data Mesh: A Tapestry of Domain-Owned Data

Data mesh empowers domain-oriented decentralized data ownership and management. It allows teams to curate and share data products across a federated data landscape. **Its core principles are:**

- **Domain-oriented decentralization**: Domains own their data and decision-making.

- **Data as a product**: Well-defined data products aligned to consumer needs.

- **Self-serve data access**: Platforms for frictionless data access across domains.

- **Federated governance**: Standards and policies implemented locally per domain.

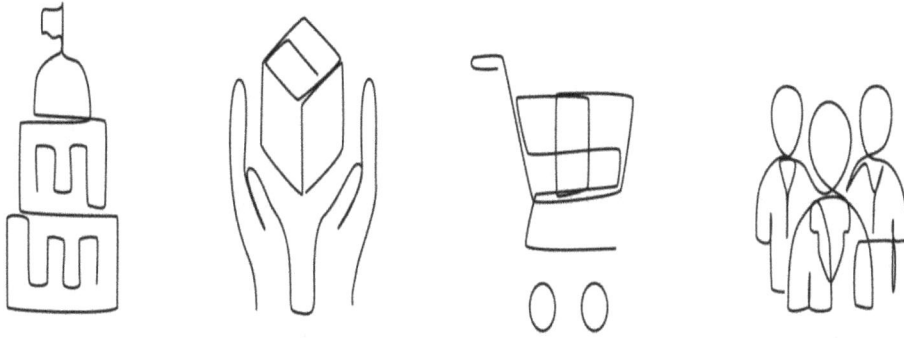

Figure 35. Data Mesh Core Principles

Enabling Data Products

Product thinking is critical to provide reliable, easy-to-consume data assets under data mesh:

- **User-focused**: Products designed for specific consumer needs.

- **Reliable**: Curated data that is accurate, timely, and complete.

- **Secure**: Robust security and access control mechanisms.

- **Actionable**: Drives insights and decision-making for consumers.

- **Accessible**: Discoverable and easy to leverage by intended users.

Figure 36. Data as a Product

Facilitating Data Marketplaces

Internal data marketplaces allow domains to discover, **collaborate on, and exchange data products:**

- **Enables collaboration**: Domains can exchange data products to drive new insights.

- **Fosters innovation**: Accelerates value creation from shared data assets.

- **Increases efficiency**: Eliminates redundant efforts for everyday data needs.

- **Requires governance**: Standards for security, privacy, and interoperability.

Figure 37. Data Marketplace

- **Example**: A bank implements a data mesh with domain-owned data products. The marketing team curates customer profile data, risk manages credit data, and finance governs financial reporting data. An internal data marketplace allows for discovering and collaborating on data to generate insights.

Lakehouse

The lakehouse architecture combines the capabilities of data warehouses and data lakes to provide a unified analytics platform. **Key capabilities include:**

- **Ingesting and processing**: Dealing with structured, semi-structured, and unstructured data from diverse sources.

- **Universal processing**: Enabling batch and real-time data processing on the same platform.

- **Supporting analytics and insights**: Using SQL queries alongside machine learning.

- **Scaling**: Providing data governance, security, reliability, and performance at scale.

- **Asset management**: Leveraging metadata to catalog data assets in the lakehouse.

By consolidating data ingestion, processing, storage, and analytics in one place, a lakehouse enables an organization to extract value from its data quickly and flexibly.

Data Fabric

Data fabric creates an integrated layer of data across the organization. With data fabric, enterprise data is woven into a single logical platform that enables holistic analytics by removing data silos, easy discovery and access to real-time data, quick integration of new data sources, **and agile data sharing across business units and teams. Some fundamental principles include:**

- **Distribution**: Federated architecture with distributed data sources.

- **Unification**: Metadata integration provides a unified view of data assets.

- **Access**: APIs and services that enable universal data access.

- **Streamlining**: Automated ETL/ELT for streamlined data pipeline orchestration.

- **Consistency**: Governance and security policies are applied uniformly.

- **Example**: A multinational company with multiple subsidiaries worldwide implements a data fabric to consolidate its data assets. The data fabric enables the company's analysts to access and analyze data from different regions and business units in real-time, enhancing collaboration and enabling data-driven strategies.

Data Architecture Trend	Key Takeaways
Centralized vs. Distributed Data	- Centralized architecture simplifies data management and ensures data consistency.
	- Distributed architecture accommodates large volumes of data and enables parallel processing.
Data Mesh	- Data mesh empowers domain teams to manage their data products, fostering a data-driven culture.
	- Reduces dependency on centralized data teams and enhances agility.
Lakehouse Architecture	- Combines data lakes' flexibility with data warehouses' structured querying capabilities.
	- Enables real-time data processing and analytics while maintaining data reliability.
Data Fabric	- Integrates and interconnects distributed data for seamless access and analysis.
	- Provides a unified view of data across various sources, promoting agility and data-driven decisions.

Table. Key Takeaways for Data Architecture Trends

DataOps: The Sherpas of Our Journey

However, one crucial question remains. How will these architectural systems be operated, maintained, and enhanced over time? Even the most advanced data platforms require careful orchestration and engineering to realize their full potential.

This brings us to DataOps, the philosophy of applying operational rigor and software engineering practices to manage data landscapes better. If the architecture is the blueprint, DataOps is the construction crew.

In Himalayan mountaineering, sherpas are experienced local guides who know every crevice and crag of a mountain. In the data and AI expedition, DataOps embodies the spirit of the sherpa—guiding, supporting, and ensuring the safety and success of the journey. DataOps introduces operational rigor, automation, and collaboration as the backbone of an ascent, enabling the navigation of data management, architecture, and strategic complexities with agility and precision.

Foundational data management components focus on policies and standards, and robust architecture provides the infrastructure. DataOps is all about execution. It enables organizations to efficiently orchestrate data pipelines at scale, taking architectures from static to dynamic.

Case Study: Gilead Sciences

Gilead Sciences faced challenges in managing and leveraging its vast data resources for research and development. The traditional centralized data warehouse was becoming a bottleneck (Informatica n.d.; Joshi & Spens 2022; Revelation 2023; Amazon Web Services 2023; Sheeran 2024; Zinsmeister 2023; Data Mesh Learning n.d.).

- **Objective**: The goal was to decentralize data management to enhance agility and innovation, allowing specialized teams to maintain their data subsets independently.

- **Implementation**: Gilead Sciences partnered with AWS to transition to a data mesh architecture that involved:

    o Establishing a centralized data lake for storing and managing diverse data types.

    o Implementing Amazon Kendra for intelligent, ML-driven data search capabilities.

    o Using AWS services for secure, scalable, and efficient data handling.

- **Results**: The data lake was developed in nine months, and the search tool in three months, far ahead of the anticipated schedule:

    o A 50 percent reduction in the time required for data management and information retrieval was achieved.

    o The new approach enabled faster and more informed R&D decision-making.

- **Conclusion**: Gilead's data mesh implementation with AWS demonstrates the successful application of decentralized data management, leading to significant efficiency gains in R&D efforts.

Case Study: Saxo Bank

To cope with its expanding global operations and enhance its data-driven decision-making capabilities, Saxo Bank embarked on a journey to overhaul its data governance framework (De Nitto 2023; Shakir 2022; DataKitchen 2019; Saripalli 2024).

- **Objective**: The primary goal was to democratize data access across the organization, breaking down silos and enabling more efficient resource allocation, decision-making, and innovation—all while maintaining stringent compliance with financial regulations.

- **Implementation**: Saxo Bank adopted a data mesh architecture emphasizing domain-oriented decentralized data ownership and management.

- **Results**: This strategy move improved customer insights, facilitated personalized services, and led to a more collaborative and innovative environment across teams:

  - **Operational efficiencies**: The shift to data mesh led to significant operational efficiencies, with increased agility and scalability in data management and analysis processes.

  - **Compliance and regulatory frameworks**: Notable achievements included enhanced compliance and strengthened regulatory frameworks facilitated by improved data governance and quality.

  - **Customer insights and decision-making**: The implementation improved customer insights, enabling more personalized services and data-driven decision-making throughout the organization.

- **Conclusion**: Saxo Bank's adoption of data mesh set a benchmark in the financial services industry for leveraging decentralized data governance to achieve operational excellence, enhance compliance, and drive innovation.

Core Capabilities of DataOps

DataOps applies DevOps philosophies like continuous integration/continuous delivery (CI/CD) to deliver data capabilities rapidly and reliably. It provides operational rigor to orchestrate data pipelines across technologies, **processes, and teams and enables:**

- **Continuous integration and delivery**: Automate the build, test, and deployment of data pipelines through version-controlled code, reducing risks and accelerating the delivery of new data capabilities.

- **Infrastructure as code**: Provision infrastructure through code to ensure consistent, reproducible environments across development, testing, and production.

- **Monitoring**: Gauge data pipeline health, dependencies, and performance through metrics and alerts to enable proactive optimization.

- **Collaboration**: Foster alignment between data engineers, developers, analytics, and ops teams to share context and domain knowledge through the delivery lifecycle.

- **Communication**: Ensure transparency on data projects across stakeholders through documentation, status reports, demos, and collaboration tools.

DataOps enables organizations to keep pace with business by rapidly translating data into insights. It emphasizes collaboration, automation, and efficient workflows that promote a culture of continuous improvement, mirroring spiritual teachings on embracing change and evolving in alignment with one's highest purpose.

By applying DataOps, organizations adopt a mindset that values agility, adaptability, and the seamless flow of information, reflecting a spiritual approach to navigating the ever-changing landscape of data and technology.

Continuous Integration and Delivery

DataOps leverages CI/CD principles from software development to industrialize and automate data pipeline delivery.

Continuous integration (CI) focuses on developer productivity and automates the integration and testing of code changes from multiple developers to provide rapid feedback on quality. For data teams, CI auto-tests new features and validations added to the pipeline codebase.

Continuous delivery (CD) focuses on releasing software frequently and reliably. CD automates build, test, and deployment steps to streamline software releases. For data teams, CD auto-deploys vetted code changes into downstream environments.

CI/CD enables code-driven automation of the data pipeline build, test, and release processes. Version control maintains integrity across pipeline revisions. With CI/CD, data teams ship features faster without compromising stability or governance.

Infrastructure as Code

DataOps uses infrastructure as code (IaC) to enable the reproducibility and consistency of data pipeline environments. Infrastructure is defined in code and version-controlled. IaC provides the following benefits:

- **Consistency**: Repeatability of environments across development, testing, and production phases.

- **Traceability**: Visibility of ongoing changes in the infrastructure.

- **Collaboration**: Collaboration between data engineers and operations teams on infrastructure design and changes.

- **Scalability**: Ability to scale resources up or down based on workload demands.

- **Version control**: Versioning infrastructure code and changes for audibility.

IaC tools like Terraform and CloudFormation are widely used to define infrastructure declaratively, enabling automation and consistency. Additionally, containerization tools like Docker offer portability of data pipelines across different environments.

Monitoring

DataOps establishes monitoring practices to ensure the data pipelines' health, performance, **and reliability. Monitoring involves:**

- **Collecting metrics**: Capturing key performance metrics and events from data pipelines.

- **Alerting**: Setting up alerts to proactively notify teams of pipeline issues or failures.

- **Visualization**: Visualizing metrics and pipeline performance to identify bottlenecks and areas for improvement.

- **Incident management**: Defining processes to handle incidents and restore data pipeline operations.

Monitoring data pipelines allows teams to react quickly to issues, optimize performance, and ensure that data pipelines meet business requirements.

Case Study: Hitachi Vantara

A renowned gaming company faced challenges in managing and optimizing its vast data resources, essential for sustaining its rapid growth and innovation (Naess 2021; BrightTALK 2021; DataKitchen 2019; Hitachi Vantara Corporation 2021, 2022).

- **Objective**: To enhance data management and operational efficiency by implementing DataOps principles, focusing on automation and scalability.

- **Implementation**: The company partnered with Hitachi Vantara to revolutionize its data management strategies. Hitachi Vantara introduced advanced monitoring tools and practices as part of the DataOps framework. These tools automated data integration processes enabled agile and scalable workflows and ensured real-time monitoring of data pipelines.

- **Results**: The DataOps initiative, bolstered by robust monitoring, led to the creation of modern, automated data pipelines that significantly improved data management efficiency. This approach facilitated the seamless handling of diverse data types and volumes, enhancing operational efficiency and enabling the company to maintain its competitive edge in the fast-paced gaming industry.

- **Conclusion**: This case study underscores the transformative potential of DataOps in modernizing data management practices through effective monitoring. Through strategic automation and workflow optimization, the gaming company achieved remarkable improvements in efficiency and agility.

Collaboration

DataOps emphasizes collaboration between data engineering, analytics, development, and operations teams. Improved cooperation leads to better contextual understanding, **shared responsibility, and rapid feedback loops. Collaboration in DataOps involves:**

- **Regular standups**: Short, daily meetings to align on progress, roadblocks, and priorities.

- **Joint design sessions**: Collaborative sessions to review architecture, requirements, and plan changes.

- **Code reviews**: Collaborative feedback and review of code changes.

- **Documentation**: Shared documentation on data pipeline designs, changes, and operational procedures.

By fostering a collaborative culture, DataOps enables teams to build and maintain high-quality data pipelines, accelerate feedback loops, and reduce silos.

Communication

Communication is crucial in DataOps to ensure alignment across stakeholders, provide transparency on project progress, **and address concerns promptly. Effective communication includes:**

- **Status reports**: Regular updates on project status, timelines, and milestones.

- **Demos**: Periodic showcases of completed features and data pipeline enhancements.

- **Collaboration tools**: Platforms for real-time communication and sharing of project-related information.

Transparent communication in DataOps ensures that all stakeholders are informed and engaged, reducing misunderstandings and enabling timely decision-making.

Case Study: Etsy

Etsy faced the challenge of regularly updating its platform without disrupting service for its vast user base (BrightTALK 2021; Naess 2021; DataKitchen 2019; Hitachi Vantara Corporation 2021, 2022).

- **Objective**: The old deployment system needed to be faster and was prone to errors, leading to potential downtimes.

- **Implementation**:

  o **Continuous deployment**: Implemented to allow frequent, seamless updates.

  o **Automated testing**: Etsy's automated testing significantly reduces the likelihood of errors in updates, but it does not guarantee that updates will be completely error-free.

- o **Feature flags**: Used for controlled feature rollouts to subsets of users for testing.

- o **Monitoring and feedback**: The system's real-time health tracking was enhanced for immediate issue detection.

- o **Communication practices**: Regular status reports, demos of new features, and the use of collaboration tools ensured all teams were informed and aligned.

- o **Results**: Implementing continuous deployment and related practices significantly improved Etsy's deployment frequency and system stability while maintaining a positive user experience. Etsy values regular communication with its buyers, which could indicate a broader commitment to regular communication with all stakeholders.

- **Conclusion**: This case illustrates the effectiveness of combining DataOps principles with robust communication strategies. Transparent and regular communication helped align teams, manage stakeholder expectations, and ensure the successful implementation of continuous deployment.

Key Takeaways

- **Definition**: DataOps integrates agile methodologies, DevOps practices, and statistical process controls into data management practices.

- **Automation is critical**: Continuous integration and continuous deployment (CI/CD) play pivotal roles in DataOps, facilitating seamless and automated data pipeline updates.

- **Collaboration boost**: DataOps emphasizes cross-team collaboration, ensuring data scientists, engineers, and operations teams work in harmony.

- **Enhanced monitoring**: Real-time monitoring and rapid feedback loops in DataOps ensure data pipelines remain robust, efficient, and error-free.

The Emergence of Generative Intelligence and Beyond

The relentless advance of AI signifies a paradigm shift in how we approach data. GenAI and emerging technologies promise incremental improvements and a redefinition of the data landscape. New technologies are poised to take on complex tasks, streamline data governance, and unlock innovative data architectures.

These advancements mean data management could become more intuitive and adaptive, platforms more efficient and automated, and operations revolutionized with intelligent automation. **It is an exciting horizon that beckons with heightened efficiency and novel discoveries in the realm of data:**

- **Reimagining data management**: AI is already automating mundane governance tasks like classification, quality checks, and issue remediation. As AI grows more powerful, entire components can be reimagined and elevated. AI systems will harmonize master data by continuously learning matching logic. Metadata can be auto-tagged via natural language processing. GenAI will assist data stewards by responding to governance queries in domain terms.

- **Rearchitecting data platforms**: GenAI will enable next-gen data architectures, and AI builders can translate functional requirements into optimized data infrastructure designs. Complex data fabric integrations can be auto-coded, and query engines may dynamically rewrite SQL for performance.

- **Reinventing operations**: DataOps will leverage AI more deeply for intelligent automation. Pipelines can be auto-generated from user stories. ML models will monitor systems and auto-remediate errors. Bots can handle tier 1 support tasks.

- **Collaboration**: AI agents can provide context, documentation, and mentorship for data teams.

The future of DataOps is automation and intelligence augmenting human creativity.

Case Study: Fidelity

Fidelity Investments leverages AI models across its financial services operations, emphasizing robust governance and effective communication to ensure successful implementation and operation (ModelOp n.d.; Walker 2023; Fidelity n.d.; Process. st n.d.).

- **Objective:** To implement AI models across operations focusing on effective governance, accountability, and transparent communication.

- **Implementation:**

  - **Monitoring:** Fidelity deployed AI orchestration platforms to monitor models in production. According to a case study by ModelOp, Fidelity uses these platforms to deliver robust, value-generating models at speed and keep them operationally efficient.

  - **Governance:** Fidelity emphasized AI governance as a crucial aspect of their AI operations. It implemented governance frameworks to manage risks, ensure compliance, and maintain the auditability of AI applications.

  - **Communication practices:** Fidelity ensured regular communication through:

  - **Status reports:** Updates on AI model performance and operational status.

  - **Demos:** Showcasing new AI capabilities and improvements to stakeholders.

  - **Collaboration tools:** Using platforms like Slack and Microsoft Teams for real-time updates and issue tracking.

- **Results**:

  - **Monitoring**: Implementing AI orchestration platforms enabled Fidelity to monitor hundreds of AI models efficiently, improving operational efficiency and quick issue resolution.

- o **Governance**: Fidelity's commitment to AI model governance ensured that risks were managed effectively, compliance was maintained, and AI applications were auditable. This governance framework supported the organization's scalable and reliable use of AI.

- o **Communication**: Effective communication practices ensured that all stakeholders were informed about AI initiatives, reducing misunderstandings and enabling timely decision-making. Regular updates and transparent reporting helped align teams and manage stakeholder expectations.

- **Conclusion**: Fidelity's adoption of AI models in operations underscores the need for robust governance and effective communication to ensure accountability and reliability in financial services. Their focus on monitoring, management, and transparent communication practices highlights the importance of managing AI risks and ensuring the ethical and compliant use of emerging technologies.

Case Study: AstraZeneca

AstraZeneca has integrated AI into various drug development processes, focusing on improving efficiency, accuracy, and governance (Integra IT 2024; Kumar & Fioretti 2023; Alsumidaie 2024; BioSpace 2021; AstraZeneca n.d.; Brown 2023; DeMello 2024).

- **Objective**: To enhance drug development processes by integrating AI, significantly improving efficiency and effectiveness.

- **Implementation**:

  - o **Analysis and automation**: Utilized AI systems for drug development tasks such as medical image analysis, identifying potential drug targets, and predictive modeling for disease mechanisms. This included partnerships with AI specialists and using advanced imaging modalities to gather new disease insights.

  - o **AI ethics and governance**: Introduced ethical principles and AI governance within the company's operational framework to ensure responsible AI use, including guidelines on data usage, model transparency, and accountability. AstraZeneca implemented an AI Governance Framework, Risk Framework, and Playbook to ensure ethical and responsible use of AI.

- **Results**:

  - o **Streamlining drug development**: AI implementation resulted in more streamlined drug development processes, which reduced the time required for data analysis and decision-making.

  - o **Enhanced accuracy**: AI-driven predictive modeling improved the accuracy of drug efficacy predictions, leading to better-targeted clinical trials and faster identification of promising drug candidates.

  - o **Cost savings**: Achieved significant cost savings by automating routine tasks and improved resource allocation.

- **Improved management**: Enhanced management through better data governance and AI ethics, ensuring compliance and reducing risks.

- **Conclusion**: AstraZeneca has made significant strides in employing AI systems to enhance its drug development process. The focus on AI governance, ethical AI principles, and the automation of various tasks has positioned the company as a leading example of AI integration in the pharmaceutical industry.

Tailoring Data Strategies: From Startups to Enterprises

Organizations of varying sizes—from nimble startups to expansive enterprises—can customize and apply the data management and AI integration frameworks presented in this chapter.

- **Smaller organizations or startups**: The focus is on agility and establishing a robust yet flexible data foundation. This groundwork is particularly critical because it can significantly affect scalability and the ability to respond to rapid changes in the market. Startups should emphasize building a data-literate culture from the onset, ensuring that data-driven decision-making grows as the company grows. Such organizations should look towards practical applications of their burgeoning data capabilities, potentially leveraging lightweight and adaptable tools that can evolve with their business.

- **Mid-sized organizations**: These are uniquely positioned to harness the advantages of both small-scale agility and larger-scale stability. They can balance centralizing and decentralizing data efforts, ensuring governance while empowering domain-specific teams with enough autonomy to innovate.

- **Large enterprises**: These can look to refine and integrate complex data systems and structures via DataOps, a data-driven culture, and AI integration to provide a pathway to streamline processes and foster a pervasive, sophisticated data-centric ethos.

Regardless of size or structure, each organization must personalize the application of these frameworks to their unique situations.

Conclusion

This chapter has covered the groundwork, not the pinnacle. The actual summit lies in applying these principles dynamically within an organization's unique context.

Consider this chapter as setting up a base camp equipped with all the necessary tools and supplies for a climb. The next chapter will focus on building the climbing team—the stage at which collaboration, team dynamics, and leadership are crucial. The goal is to assemble a skilled and cohesive team that can leverage the established base camp to continue scaling new heights and gain access to the vistas that data and AI present in the digital age.

CHAPTER 11

Building a Climbing Team

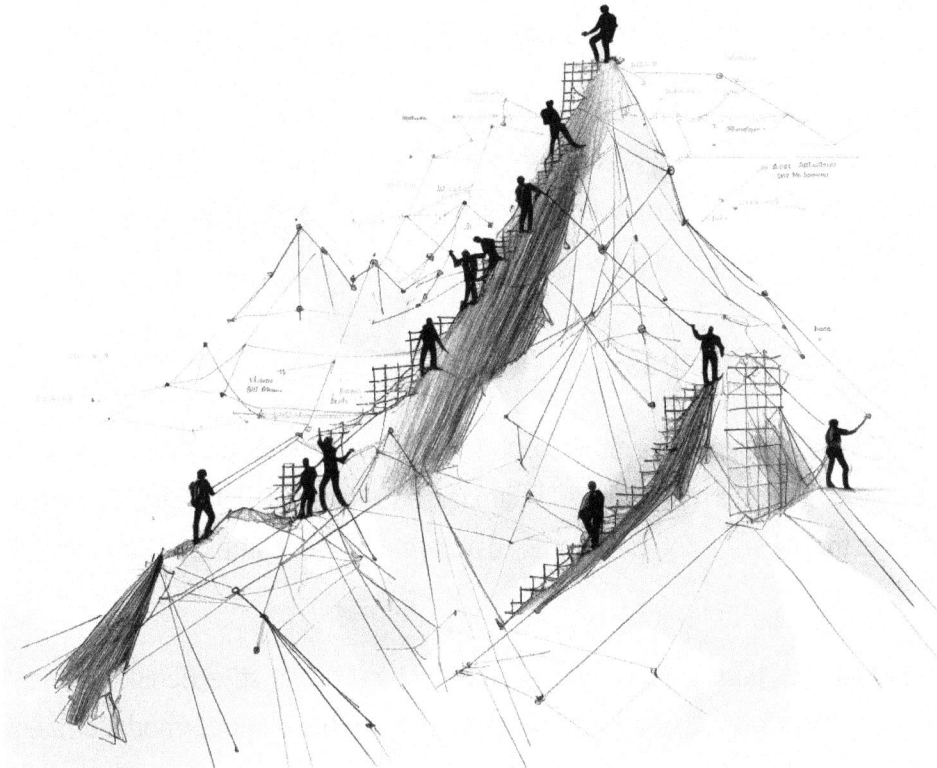

Just like a team of climbers meticulously planning their ascent, organizations must build a team adept at navigating the steep inclines of the data and AI landscape. This means attracting and nurturing talent rich in data insight and AI expertise, which is essential to fostering an environment where data is an asset and at the core of decision-making.

Embracing audacious goals is essential. This mindset encourages moving beyond perceived limitations to embody the belief that no peak is too high to conquer. There must be a shared vision and purpose that fosters a collective consciousness. Such unity, reminiscent of climbers moving together toward the summit, ensures that the journey to create a data-driven culture is marked by seamless collaboration and a deep sense of shared destiny. As explored in *The Buddha and the Badass*, uniting under a shared vision amplifies creative potential and problem-solving capabilities. Operating as one, creative potential and collective problem-solving capabilities are amplified.

Figure 38. Navigating the Learning Curve: From Novice to Expert

Leverage emotional and cultural intelligence to deepen leadership strategies. Understanding and respecting a team's diverse emotional and cultural backgrounds fosters a supportive and inclusive data-driven culture. Utilize storytelling to connect on a human level, make data and AI insights compelling and relatable, and reinforce the organization's shared mission.

Charting the Terrain for Success

Just as mountain expeditions require a precise balance between centralized planning and the climbers' autonomy, organizational models necessitate a similar equilibrium. In the vast information landscape, companies must find the sweet spot between a centralized base camp and the flexibility of individual climbing teams.

- **Base camp and summit teams**: The center of excellence (CoE) is an expedition's base camp, a central point for governance, strategy, and standardization. From this blended-model strategic hub, domain-specific teams—like summit teams—can begin their journey with tailored solutions to the unique conditions of their respective challenges. With the standards and practices established at the base camp, they have the autonomy to innovate yet remain anchored to the core mission.

- **Mapping data routes**: Treat an organization's data capabilities as routes mapped out for various climbs. Each route (data product) is designed to meet the needs of specific climbers (users), offering them a path to insight acquisition that enables them to conquer their unique business peaks.

- **Sharing supplies**: Foster an ethos of resource-sharing among climbing teams by facilitating internal data sharing. Encourage the exchange of experiences, insights, and tools—just as climbers would share supplies and knowledge to ensure the success of every member's ascent.

- **Clear signals**: Like calls between climbers that echo across the mountain, maintain clear lines of communication. This ensures that every team member, regardless of their altitude or path, can navigate without the risk of data silos or information voids and contribute to the overall mission.

The Mountaineering Council

Setting up a data center of excellence (CoE) is like establishing a council of seasoned mountaineers, each of whom can expertly guide the organization's journey. **There should be a clear team structure so that everyone's role in the expedition is defined:**

- **Leadership**: The CoE is led by visionary yet pragmatic individuals who map the organization's course through new data territories. The head of the CoE, akin to a chief expedition leader, not only charts the path but inspires the team with a compelling vision.

- **Operational roles**: Like climbing specialists, each operational team member has a pivotal role. Data engineers lay the ropes and ladders, ensuring the data infrastructure is secure and scalable. Data scientists read the weather patterns and terrain, providing insights that guide business decisions. Data analysts scout ahead, identifying new opportunities and paths through the analysis.

- **Support roles**: Essential to any expedition is the support crew that prepares climbers with the necessary skills and tools for their journey. Training and development leads mentor new climbers, while data tooling and infrastructure specialists ensure that the best equipment is always available.

The Climbing Mandate

Adopting a constant learning and adaptability mindset is essential in the ever-changing technological landscape. Fostering an environment where teams are encouraged to remain agile and ready to evolve with new technological advancements and shifts in the business landscape ensures survival and thriving success. This adaptability, rooted in a shared commitment to growth and innovation, will carry a team through uncertain terrains to new heights of achievement.

The CoE's responsibilities are clear, from strategic planning to implementing data initiatives. Like a climbing mandate, they ensure that every ascent is well-planned, equipped, and executed.

- **Unpredictable elements**: Every climb faces unpredictable elements, from sudden storms to unseen crevasses. The CoE must navigate cultural resistance, data silos, and the constant evolution of technology, which is akin to the shifting and treacherous terrains of high mountains.

- **Success factors**: A CoE's success is measured by its ability to reach the summit and accomplish the mission. This requires unwavering executive support, a commitment to continuous learning and adaptation, and a collaborative spirit that unites all climbers in the organization's purpose.

Fostering a Data-Driven Culture

To become a data-driven organization, leaders must drive a cultural shift that prioritizes data-driven decision-making and embraces data-driven initiatives.

- **Secure executive sponsorship**: Gain the support and commitment of top leadership to champion data initiatives throughout the organization. Executive sponsorship provides the necessary resources and backing to drive data-driven transformation.

- **Tie initiatives to business value**: Align data initiatives with strategic business goals and emphasize how data-driven decisions contribute to tangible outcomes, such as increased revenue, reduced costs, and improved customer experiences.

- **Incentivize data-driven decisions**: Create incentives and rewards for employees who embrace data-driven decision-making. This encourages a data-first mindset across the organization.

- **Communicate successes and lessons learned**: Celebrate successes that result from data initiatives and openly share lessons learned from successes and challenges. This promotes a culture of learning and continuous improvement.

- **Promote data literacy**: Building data literacy across the organization to enable data-driven decision-making at all levels.

- **Provide training on data fundamentals**: Comprehensive training programs on data fundamentals, data analysis, and data visualization ensure that employees have the skills to interpret and leverage data effectively.

- **Enable access to data via self-service analytics**: Empower employees with self-service analytics tools that allow them to access and analyze data independently. This democratizes data access and empowers employees to make data-driven decisions.

- **Use clear communications**: Communicate data insights and findings in a direct and accessible manner, avoiding technical jargon. This ensures that all stakeholders understand the value of data-driven decisions.

- **Build understanding incrementally**: Start with simple data concepts and gradually build understanding and complexity. This approach helps individuals feel more comfortable with data analytics and fosters confidence in their decision-making.

- **Encourage curiosity on data impact**: Promote a culture of curiosity around data and its potential impact on business outcomes. Encourage employees to explore data and propose data-driven solutions.

- **Leadership commitment and evangelizing success**: Senior leaders must actively champion the data-driven culture and demonstrate its impact through their actions and decision-making.

Charting with Informed Insight

Language has been the compass by which humanity navigated the world's complexities, sharing knowledge from generation to generation. In *The Code of the Extraordinary Mind*, Vishen Lakhiani discusses how language profoundly shapes our reality. Consider the Himba tribe in Namibia, whose language does not distinguish blue from green. This linguistic nuance raises profound questions about perception and awareness. Does it become invisible without a concept or label for an object?

This anthropological curiosity echoes a phenomenon known as the "Columbus effect," named after an apocryphal tale in which Native Americans supposedly could not initially see Columbus' ships because they were so outside their experience. Although this story is more myth than fact, it poignantly illustrates that recognizing new possibilities can be challenging without previous understanding or context.

This notion is pertinent when we scale the modern business landscape. Today's data is uncharted territory, rich with hidden insights akin to undiscovered lands. However, these insights can only illuminate a path forward if recognized and understood—which is where data literacy becomes critical.

Data literacy is an individual skill and a collective language that allows an organization to see and navigate. It enables employees to ask the right questions, identify patterns, and extract actionable information. With data literacy, a workforce can access maps and read them, empowering them to actively participate in an organization's journey to innovation and decision-making.

However, a gap in this literacy is like standing on a cloudy peak; the landscape beneath is assumed, but its features cannot be discerned. Modern businesses can equip themselves with the most advanced analytical tools, akin to the finest climbing gear, but still get stuck at base camp without the ability to move forward through the data terrain.

Unlike the Himba, the Russian language has separate words for light blue (голубой [goluboy]) and dark blue (синий [siniy]). This distinction subtly shapes perception and attention—similar to how businesses view different data sets. By establishing a vocabulary for various data concepts, organizations can enhance their ability to discern and act upon nuances that might be overlooked.

To bridge the language gap, organizations must prioritize fostering data literacy as an essential skill across their teams. Providing the proper training and resources can empower employees with the language of data, just as learning new terms allows us to perceive what was always before us in a new light. By nurturing a data-literate culture, companies can fully harness the power of their data. In this way, the journey of data literacy is more than learning a skill—it is about expanding the collective vision. By investing in data literacy, organizations are empowered not just to look but to see, climb, and forge new paths, and not just exist in a data-centric world but to thrive and lead it.

Storytelling as a Way to Map the Journey

In the expansive terrain of data-led transformation, storytelling emerges as a tool and a guiding light, distinguishing data scientists from their statistician predecessors. The art of turning data into a narrative propels organizations forward, transforming the abstract and often impenetrable into the tangible and relatable. This skill does more than communicate; it transports listeners on a journey, turning data points into waypoints on a map of discovery and decision.

Once upon a time, a select few navigated the realm of data, armed with their compasses of statistical insight. However, the emergence of big data and the pressing need for actionable intelligence have heralded the rise of a new navigator: the data scientist. These modern explorers are adept in the science of data and the art of storytelling, understanding that numbers gain meaning only when woven into a narrative that resonates across an organization's entire expanse.

The dawn of GenAI marks a new chapter in this exploration, democratizing the tools of storytelling and breaking down the walls of the data silos. Now, anyone—from executives to frontline workers—can embark on voyages of discovery armed with AI as their guide. This shift represents a seismic change in how insights are generated and shared, inviting all to contribute to the collective journey.

Consider the endeavors of news organizations like the BBC and the *Guardian*, which navigated the vast sea of big data to bring stories to life. The BBC, for example, uses big data analytics to understand audience preferences and behaviors, allowing them to tailor content and deliver personalized experiences (Barr 2019; Webb 2020; BBC Datalab n.d.; Inrupt n.d.; Bounegru n.d.; Kent 2012). Their Visual and Data Journalism team creates interactive graphics and data-driven stories that engage readers in new and innovative ways. Similarly, the *Guardian* established a Data Projects team that analyzes large datasets to uncover hidden stories, such as their award-winning investigation into the US National Security Agency's surveillance activities (Benton 2010; Rogers 2021; Pulitzer Prizes n.d.; Lakhiani 2016). By assembling crews of designers and data scientists, these news organizations charted courses through complex information, making it accessible and engaging to a global audience. Their success lies not just in the skill of navigation but in their ability to tell stories that captivate and inform.

GenAI amplifies this capability, offering a sextant to chart the unseen, making discovery, analysis, and presentation efficient and intuitive. This fusion of technology and narrative empowers every member of an organization to tell their stories, turning data into a shared language that informs decisions, drives innovation, and fosters a truly data-driven culture.

Through the lens of GenAI, storytelling becomes a collective endeavor, a shared responsibility that transcends departmental boundaries. It is a force that drives not just understanding but action, bridging the chasm between the raw material of data and the impactful outcomes it can achieve. In this light, storytelling is a tool and a compass, guiding organizations through the uncharted territories of data toward transformation and success. Every charted course and every shared story contributes to the collective journey that navigates the present and shapes the future.

Navigating Data Expedition Roles

As with any well-planned expedition, the success of a data-driven journey hinges on the clarity of roles within the team, the responsibilities assigned to each member, and the continuous development of skills necessary to navigate the challenges ahead.

- **Roles and responsibilities**: A good team is diverse, with each member's specialized tools and knowledge crucial. From strategists who map the journey's objectives to engineers and scientists who interpret the terrain and analysts who scout ahead, every role is vital. Just as an expedition leader delineates responsibilities to ensure the safety and efficiency of the journey, organizations must clearly define data roles to avoid ambiguity and ensure success.

- **Central vs. domain teams**: Imagine central data teams as base camp personnel who provide overarching support, governance, and resources, while domain-specific teams operate like individual expedition members, adapting strategies to their unique environmental conditions. This structure supports centralized oversight and the flexibility for specialized teams to innovate within their domains.

- **Building data skills and capabilities**: Comprehensive data analytics, visualization, and AI training introduces team members to the tools and techniques essential for navigating the data wilderness. Workshops and hackathons are the equivalent of orienteering exercises, enhancing the team's ability to find innovative paths through complex data challenges.

- **Self-service analytics and AI proficiency**: Democratizing data through self-service tools equips every expedition member with the means to contribute insights, much like giving each climber navigational tools. This empowerment, combined with specialized training in AI, prepares the team to tackle even the most daunting data peaks and ensures that the organization can move forward confidently—even in the face of new and unforeseen challenges.

Cultural Transformation

Transitioning to a data-driven culture is a journey encompassing more than mere tools and processes—it is about the collective spirit of exploration, understanding the landscape, and moving together toward a shared vision.

- **Mapping the terrain**: Just as explorers must understand the terrain to navigate it, leaders must grasp the root causes of resistance to change. Is it fear of the unknown, a sense of vulnerability, or a lack of understanding? Engage in dialogue—imagine gathering around a campfire—to share stories, listen to concerns, and foster mutual understanding and respect.

- **Leadership**: Every expedition needs its guiding stars, leaders who provide direction and inspiration. Visible commitment from senior leadership acts as a beacon, encouraging everyone to align their steps with the journey ahead. Data champions emerge as the scouts and navigators, bridging gaps and guiding teams through the data wilderness.

- **Transparent communication**: Clear and open communication ensures that every voice is heard and acknowledged. You articulate the "why" behind the journey, and maintaining open channels for feedback, and concerns ensure that everyone understands and is aligned with the mission.

- **Inclusive decision-making**: Involving everyone in mapping the route ensures the expedition is a collective effort. Whether selecting the right tools or designing workflows, every team member's input should be valued, like choosing the best path up a challenging peak.

- **Taking uncharted paths**: Empowering individuals with self-service analytics tools equips them with what they need to explore uncharted paths and contribute unique insights. Cultivating a culture of curiosity and exploration fuels innovation and discovery.

- **Cultural reinforcement**: Integrating data into daily routines is how the new culture takes root and grows. Recognition and rewards for data-driven achievements are the celebrations that strengthen the team's resolve and commitment to the journey.

- **Continuous support**: Supporting each other is the lifeline that ensures the expedition's success. Providing ongoing resources, training, and feedback mechanisms keeps everyone aligned and moving forward, adapting to new challenges.

This approach makes the shift towards a data-driven culture more relatable by framing it as a collective journey. It highlights the importance of leadership, engagement, education, and continuous support. It reminds us that this transition is not just a technical shift but a profound cultural transformation that requires everyone's participation and commitment.

Seeing Over the Horizon

In the evolving landscape of data and AI, organizations must adapt to present changes and anticipate tomorrow's roles. The roles we created today may become tomorrow's standard.

- **Prompt engineers**: As AI becomes more conversational and interactive, prompt engineers are emerging as the linguists of AI, crafting the language that dictates behavior and output. They are the translators between human intent and AI action.

- **AI detectives**: As AI proliferation increases, the need to understand and investigate decision-making processes will lead to AI detectives dissecting algorithms to ensure transparency, uncover biases, and safeguard ethical standards.

- **Virtual world architects**: The rise of digital realities requires the creation of immersive, functional, and engaging virtual spaces. Virtual world architects will design and govern these new realms that blend AI with virtual and augmented reality.

- **Data trust officers**: As data privacy remains paramount, Data Trust Officers will ensure that data policies are respected, data usage is transparent, and trust is maintained between organizations and their stakeholders.

- **Beyond the status quo**: Innovation thrives at the edge of the known world. To foster an ecosystem where new roles and technologies flourish, organizations must establish innovation labs and partnerships that function like exploratory outposts, scouting emerging technologies and cultivating breakthroughs that redefine the marketplace.

- **Staying atop industry trends**: It is someone's responsibility to follow the shifting winds of industry trends and emerging technologies. This vigilance is far from needless—it is the lifeblood of future-ready organizations. Leaders must immerse themselves in the currents of continuous learning, attend conventions where thought leaders gather, and participate in forums where the future is being written.

Data Organizational Models in Modern Enterprises

Data management structures vary across organizations based on several factors, including size, culture, and industry demands. This section explores different data organizational models that align with operational needs and strategic goals.

Spotify: Agile and Autonomous

- **Structure**: Spotify organizes its workforce into decentralized teams called squads, along with other team structures like chapters, tribes, and guilds. Squads are grouped into tribes, with tribes consisting of multiple squads within the same business area. Spotify utilizes chapters to promote team collaboration and innovation, with chapter leads responsible for developing people and setting salaries within a particular area. Guilds operate informally to encourage knowledge sharing, spanning the entire company (Hardy 2016; Boyd 2019; Cruth n.d.; Berg 2020).

- **Why this works**: This structure's autonomy promotes cross-functional collaboration and embeds continuous learning into the organizational culture.

Meta: Integrated Data Science Approach

- **Structure**: Meta operates a central organization model for its Core Data Science team, which partners with various teams throughout the company. While data scientists are not directly embedded within product teams, they collaborate closely with product managers, engineers, and designers across different areas to integrate data-driven insights into all stages of product development. Meta employs a hybrid approach, with centralized teams focusing on overarching strategies and embedded teams working within specific functional departments (Meta 2022; Infinite Monkey 2022).

- **Why this works**: By embedding data science expertise within product teams, Meta ensures that data-driven insights are seamlessly integrated into product innovation, enhancing the relevance and responsiveness of its products. The centralized teams provide a backbone of advanced analytics and strategic direction, ensuring consistency and high standards across the organization.

Airbnb: Academic and Exploratory

- **Structure**: Airbnb has modeled its data science team on academic structures, with different areas of specialization akin to university departments. Leadership roles are comparable to tenured professors who drive long-term research and exploration (Nasser, 2023; Christensen, Bartman, & van Bever, 2016).

- **Why this works**: This model underscores Airbnb's value in deep, foundational research and adaptability, supporting sustained innovation and thorough exploration of new data frontiers.

Netflix: Centralized Data Expertise

- **Structure**: At Netflix, centralized data teams are the hub of expertise that serves the entire organization. This model nurtures a culture where access to data is universal, and each team is empowered to make informed decisions (Mondal 2020; van Es 2022).

- **Why this works**: Centralization of data expertise ensures consistency, fosters a culture of transparency and trust, and promotes the development of specialized knowledge that benefits multiple product areas.

LinkedIn: The Hybrid Approach

- **Structure**: LinkedIn's hybrid data organizational model positions data scientists within product teams for targeted action and within a central analytics unit for overarching strategy. This dual presence ensures tight alignment with product needs while maintaining a unified data strategy (Vashishta, 2023).

- **Why this works**: The hybrid model supports consistent data strategy at the macro level and grants data scientists the proximity and autonomy needed for impact at the micro level, effectively balancing strategic cohesion with agile innovation.

Each of these models is fluid, capable of evolving as the company grows and the role of data expands. This ensures that the organizational structure continues to serve the changing needs of the business and its strategic objectives.

In the data-driven journey, fostering an environment where teams are encouraged to remain agile and ready to evolve with new technological advancements and shifts in the business landscape ensures survival and success. Such adaptability, rooted in a shared commitment to growth and innovation, will move businesses through uncertain terrains to new heights of achievement.

Conclusion

The journey toward digital transformation is as much about personal growth as organizational change. As individual team members grow and embrace a data-driven mindset, the organization evolves, moving closer to achieving innovation and efficiency. This collective progression is the essence of transforming into a data-driven entity.

The operational models discussed—centralized, decentralized, and hybrid—are different routes up the mountain, each leveraging collective strengths and strategic vision. Like a team of climbers roped together, cross-departmental collaboration ensures that every member contributes to the collective goal.

CHAPTER 12

Navigating the AI Terrain

Just as climbers navigate challenging terrains, organizations need well-defined ways to succeed in data and AI initiatives. This chapter explores the essential components of defining a robust delivery methodology tailored to the unique challenges of these projects.

Compared to traditional IT projects—characterized by well-defined objectives and clear roadmaps—data initiatives are frequently characterized by high uncertainty and the need to explore options. It is common for many such projects to become trapped in a "pilot purgatory" marked by experimental and exploratory work before scalability and integration with business processes become formidable obstacles. Factors contributing to such stagnation include unclear strategies for scaling solutions, misalignment with core business goals, integration complexities, insufficient stakeholder support, and resource limitations.

The Challenges of Data Projects

One of the key challenges organizations face is the exploratory nature of data initiatives, which is only sometimes accounted for in standard delivery methodologies. Organizations often reach a critical point where they must fully adapt their working methods to embrace this exploratory element. **Challenges arise because data projects involve:**

- High uncertainty and variability in outcomes based on the data.

- A high degree of reliance on data quality and integrity.

- There is a need for continuous exploration and hypothesis testing.

- Cross-functional collaboration between technical and business teams.

Without a tailored methodology to guide the journey, **organizations grapple with the following issues:**

- **Unclear business objectives and success metrics**: Often, organizations embark on data and AI initiatives without clearly understanding their business objectives and the metrics that will define success. Without well-defined goals, they risk wandering while unable to gauge progress or the impact of efforts.

- **Misalignment between data, analytics, and engineering teams**: Data projects involve multidisciplinary teams, including data scientists, data engineers, analysts, and business stakeholders. Misalignment can lead to confusion, inefficiencies, and disjointed solutions.

- **Difficulty transitioning experimental models into production**: It is one thing to develop promising AI models and data solutions in controlled environments; it is another to deploy them successfully into production systems. Transitioning from experimentation to production often presents significant challenges, leading to delays and missed opportunities.

- **Lack of continuous monitoring and improvement**: Data and AI solutions are dynamic, requiring constant monitoring, feedback, and improvement to stay relevant and practical. Organizations must keep up or risk being left behind as the data landscape evolves.

- **Inability to demonstrate tangible ROI to justify investments**: Securing funding and resources for data and AI projects often hinges on establishing a tangible ROI. Organizations may need a clear path to showcase the value of data initiatives to gain the necessary support.

To overcome these pitfalls, a new methodology designed specifically for the world of data must be embraced.

The Way of Exploration

The Shu Ha Ri model, covered in Chapter 2, provides valuable insights into this adaptation process. The Shu phase represents strict adherence to established methodologies, which may work for well-defined IT projects. However, the need for a more flexible and adaptive approach becomes apparent as data projects progress into the experimental Ha phase. This realization marks the Ri phase, where organizations must transcend their current work, fully embrace data projects' exploratory nature, and achieve innovation.

Embracing the Ri phase involves adopting principles that allow organizations to thrive in the exploratory landscape of data projects:

- **Flexibility and iteration**: Embrace continuous learning and adaptability, allowing feedback from each exploration loop to inform the next steps.

- **Agile and adaptive methodologies**: Traditional waterfall methodologies may not be suitable for data projects. Agile methods offer more flexibility, allowing teams to adapt to changing data realities.

- **Experimentation and prototyping**: Cultivate an innovation culture. Test multiple hypotheses, learn from failures, and refine and improve solutions.

- **Data quality and governance**: Prioritize data quality and establish robust governance practices to ensure data integrity and reliability.

- **Cross-functional collaboration**: Effective data exploration requires coordination between diverse teams. Encourage cross-functional teams to foster different perspectives and enhance problem-solving capabilities.

- **Risk management**: Proactively identify and manage hazards by anticipating potential roadblocks and planning mitigation strategies.

- **Knowledge sharing**: Emphasize a learning culture where insights and information are shared across teams for continuous improvement.

A Tailored Methodology for Data and AI Initiatives

This section presents a data expedition methodology that is a phased approach allowing for the iterative development and refinement of solutions, **from proof of concept to minimum viable product to a full-scale production system:**

- **Proof of concept (POC)**: Validates technical feasibility during design and development.

- **Minimum viable product (MVP)**: Contains core features for pilot testing during development.

- **Production-Ready Solution**: Finalized and launched during the operational phase based on MVP feedback.

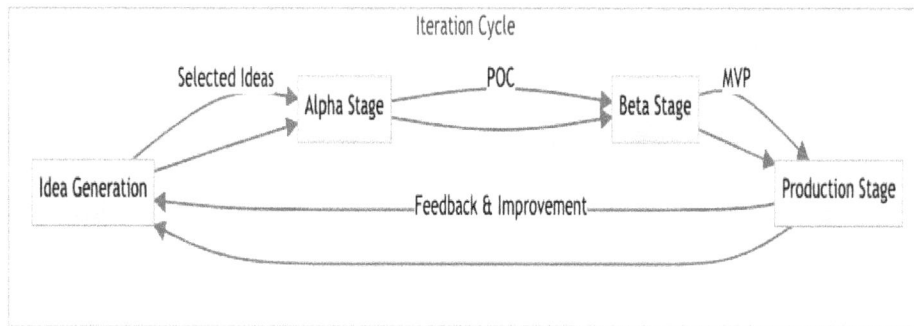

Figure 39. Delivery Methodology for Data and AI Initiatives

By incorporating rapid iteration with user validation, this methodology balances ambition with practicality to drive maximum business value.

Discover: Opportunity Identification

The discovery phase is the starting point, focusing on understanding an organization's data assets, identifying gaps, and pinpointing opportunities where AI can drive value. This phase is all about exploration and alignment. By diving deeply into an organization's data landscape and business priorities, the groundwork for subsequent stages is laid. Collaborative workshops, SWOT analysis of data assets, and feasibility studies are vital activities during this phase. The outcome will be a clear roadmap of AI projects that align with business goals and data capabilities.

- **Objective**: Understand the organization's data assets, identify gaps, and spot opportunities where AI can drive value.

- **Team**: Data strategists, business analysts, domain experts, and data engineers.

  o **Data strategists**: Identify data assets and potential AI applications.

  o **Business analysts**: Align AI opportunities with business goals.

  o **Domain experts**: Provide industry-specific insights.

  o **Data engineers**: Assess data quality and availability.

- **Checkpoints**: Complete data inventory and AI opportunity roadmap.

- **Guiding questions**:

  o What data assets do we possess?

  o Where can AI add the most value to our current operations?

- **Best practices**: Collaborative workshops, SWOT analysis of data assets.

- **Common challenges**: Data silos, lack of clarity on business objectives.

- **Key activities**:

    o Data inventory and quality assessment.

    o AI opportunity workshops with business units.

    o Feasibility studies considering data availability and quality.

- **Outcome**: A roadmap of AI projects aligned with business goals and data capabilities.

Design: Data Preparation and AI Blueprinting

This is where the rubber meets the road. Data engineers work diligently to integrate and preprocess data sources, ensuring they are ready for AI modeling. Concurrently, data scientists and solution architects collaborate to design the most appropriate AI model architectures, considering business objectives and the nature of the data. The outcome of this phase is straightforward: an AI model design and a dataset primed for training.

- **Objective**: Ensure data readiness and design an appropriate AI-model architecture.

- **Team**: Data engineers, data scientists, and solution architects.

    o **Data engineers**: Prepare and integrate data sources.

    o **Data scientists**: Design AI model architectures.

    o **Solution architects**: Blueprint the overall data and AI pipeline.

- **Checkpoints**: Data readiness, AI model design approval.

- **Guiding questions**:

    o Is our data of sufficient quality for AI modeling?

    o What AI model architectures best suit our needs?

- **Best practices**: Iterative data preprocessing, modular AI design.

- **Common challenges**: Inconsistent data, choosing the wrong model architecture.

- **Key activities**:

    o Data preprocessing and integration.

    o AI model architecture design.

    o Data and AI pipeline blueprinting.

- **Outcome**: A straightforward AI model design and a ready-to-use dataset.

Develop: Model Training and Validation

This is the heart of the AI journey. Data scientists train AI models, fine-tuning them to achieve the best performance. They validate them using separate datasets to ensure accuracy and reliability. During this phase, feedback from domain experts is invaluable, ensuring that the AI models align with industry-specific requirements. The result is a validated AI model ready for deployment and integration into business processes.

- **Objective**: Achieve a trained AI model that meets predefined performance metrics.

- **Team**: Data scientists, ML engineers, domain experts.

  - **Data scientists**: Train and validate AI models.

  - **ML engineers**: Optimize model training pipelines.

  - **Domain experts**: Provide feedback on model outputs.

- **Checkpoints**: Model performance metrics and validation results.

- **Guiding questions**:

  - How well is our model performing against predefined metrics?

  - Are there any biases in the model predictions?

- **Best practices**: Cross-validation, hyperparameter tuning.

- **Common challenges**: Overfitting, lack of computational resources.

- **Key activities**:

  - Model training and hyperparameter tuning.

  - Model validation using a separate dataset.

  - Feedback collection from domain experts.

- **Outcome**: A validated AI model ready for deployment.

Operationalize: Deployment and Integration

This is the phase where the AI solution comes to life in the real world. ML engineers and DevOps teams collaborate to ensure the model is deployed smoothly, is scaled as needed, and remains robust in a live environment. IT teams play a crucial role in integrating AI models with existing business applications, ensuring that the predictions and insights are readily available to decision-makers. The outcome of this phase is an AI solution that's not just operational but also one that delivers tangible business value in real-time.

- **Objective**: Achieve a seamless deployment of the AI model and ensure integration with business processes.

- **Team**: ML engineers, DevOps, IT teams.

  - **ML engineers**: Prepare models for deployment.

  - **DevOps**: Ensure seamless model deployment and scaling.

  - **IT teams**: Integrate AI models with existing systems.

- **Checkpoints**: Successful model deployment and integration tests.

- **Guiding questions**:

  - How will the model be monitored during production?

  - Are there any latency requirements for model predictions?

- **Best practices**: Continuous integration and deployment, monitoring tools.

- **Common challenges**: Model drift and integration issues with legacy systems.

- **Key activities**:

  - Model deployment to production servers.

  - Integration of AI models with business applications.

  - Initial monitoring and performance tracking.

- **Outcome**: The AI model is fully operational and integrated, delivering real-time predictions.

Evolve: Monitoring and Continuous Learning

AI models must adapt as the business environment and data patterns change. This phase is characterized by constant monitoring, feedback loops, and periodic retraining to ensure models remain relevant and accurate. Data scientists closely monitor model performance, ensuring that any drift in data patterns is accounted for. Business analysts measure the ongoing impact of the AI models, ensuring they continue to drive business value. The goal is an AI solution that delivers consistent value and evolves and grows with the organization.

- **Objective**: Ensure the AI model remains accurate and relevant over time.

- **Team**: Data scientists, ML engineers, and business analysts.

  - **Data scientists**: Monitor model performance and biases.

  - **ML engineers**: Retrain models with fresh data.

- o **Business analysts**: Measure the business impact of AI models.

- **Checkpoints**: Model retraining schedules and performance reports.

- **Guiding questions**:

  - o How has the data landscape changed since the last model training?

  - o Do you know if the models are still aligned with business objectives?

- **Best practices**: Feedback loops, periodic model audits.

- **Common challenges**: Changing data distributions, model staleness.

- **Key activities**:

  - o Continuous monitoring of model predictions.

  - o Periodic retraining of the model with new data.

  - o Business impact analysis of AI predictions.

- **Outcome**: An AI model that continuously learns and evolves, maximizing business value.

Let us examine real-world case studies from industry giants to better understand this methodology's practical application.

Case Study: Dynamic Pricing Model

- **Discover**: Recognizing the fluctuating demand for properties influenced by location, season, and local events, Airbnb began exploring dynamic pricing to provide hosts with data-driven pricing suggestions to optimize occupancy and revenue.

- **Design**: Airbnb's engineering and data science teams developed a sophisticated dynamic pricing system that leverages machine learning to analyze various factors, including property characteristics, booking trends, and market dynamics, to suggest optimal prices to hosts.

- **Develop**: The ML models underpinning Airbnb's dynamic pricing system are refined using an extensive dataset encompassing historical booking data, property features, and market conditions. These algorithmic models adapt to changing market behaviors and preferences, ensuring pricing suggestions remain competitive and tailored to individual listings.

- **Operationalize**: Airbnb introduced the Price Tips and Smart Pricing tools, which allow hosts to access dynamic pricing suggestions. These tools offer a range of prices based on the calculated likelihood of booking, helping hosts make informed pricing decisions.

- **Evolve**: Airbnb continuously optimizes its dynamic pricing model in response to new data, host feedback, and evolving market conditions. This iterative process ensures that the platform's pricing suggestions remain relevant and effective in maximizing occupancy and revenue for hosts.

Airbnb's dynamic pricing model represents a critical component of its strategy to enhance platform usability for hosts and improve the overall guest experience. By leveraging advanced ML techniques and a comprehensive dataset, Airbnb has successfully created a dynamic pricing system that supports its mission of creating a world where anyone can belong anywhere (Owen 2021; ProjectPro 2024; Sims, Ameen, & Bauer 2019; Hivelr 2024; Summers 2024).

Case Study: An Agile Approach to Personalization

- **Discover**: Spotify recognized the essential role of personalization in enhancing the user experience, realizing that vast selection options could overwhelm users. By leveraging big data and analytics, Spotify set out to develop a feature that could deliver the personal touch of a mixtape made by a close friend to enhance user engagement through tailored music recommendations.

- **Design**: Spotify's team gained insights into individual listening habits by analyzing extensive user data. They employed collaborative filtering and matrix factorization techniques to design a system capable of identifying and predicting user preferences, laying the groundwork for highly personalized music recommendations.

- **Develop**: Spotify's algorithms were crafted to reflect the individual's listening history alongside the collective patterns of all users. This Discover Weekly feature underwent iterative testing and refinement, allowing Spotify to gather continuous feedback and make enhancements, ensuring recommendations remained relevant and engaging.

- **Operationalize**: With the algorithms fine-tuned, Spotify introduced Discover Weekly to its user base. This innovative feature automatically generates a new weekly playlist for each user, integrating seamlessly within the platform and marking a significant advancement in music personalization.

- **Evolve**: Discover Weekly is an ever-evolving Spotify feature utilizing ML models that adjust to user feedback. This allows the continual refinement of recommendations, ensuring they align with the latest listening trends and user behaviors. This maintains the platform's competitiveness and relevance in the digital music landscape.

Spotify's Discover Weekly exemplifies how agile workflows, combined with data and AI, can successfully scale personalized experiences to millions of users. This case study showcases an iterative development process informed by user behavior and trends, emphasizing the importance of flexibility and adaptability in digital innovation (Mulkers, 2023; Kaushik, 2024; Murphy, 2023; Pragmatic Institute, n.d.).

The Data Expedition Methodology: Guiding the Journey to Success

The data expedition methodology provides organizations with a map and compass to successfully navigate data mesh, analytics, **and AI. This methodology comprises nine interlinked components that cover the entire lifecycle of data initiatives:**

1. **Staged approach**: Ensures a structured progression from initial concept to full-scale implementation.

2. **Adapting for data and AI**: Tailor's methodologies specifically for data and AI projects.

3. **Embracing exploration, from Shu to Ri**: Encourages continuous learning and mastery.

4. **Aligning with data mesh**: Integrates seamlessly with data mesh principles for decentralized data management.

5. **Climbing with agility**: Promotes agile methodologies to remain flexible and responsive.

6. **Putting the user first**: Prioritizes user needs and feedback throughout the project.

7. **Streamlined operations**: Focuses on operational efficiency and effectiveness.

8. **Cloud computing, the digital sherpa**: Leverages cloud technologies to support and enhance data initiatives.

9. **Measuring outcomes, the data expedition compass**: Utilizes metrics and KPIs to track progress and success.

1. Staged Approach

The journey is divided into iterative stages: discover, design, develop, operationalize, and evolve. Each stage represents a base camp on the path to the summit.

- **Discover**: Expedition leaders scout the landscape to identify challenges and opportunities, surveying the terrain and noting potential pathways and obstacles. The discovery stage is critical for understanding objectives and charting the course ahead.

- **Design**: Involves detailed planning and preparation before beginning the data journey. The route is mapped out, gear is checked, and success metrics are defined. A comprehensive design provides the blueprint for a successful expedition.

- **Development**: The data climbing team builds bridges, ladders, and ropes to navigate rugged ravines and cliffs. Data pipelines and AI models are constructed iteratively and tested for readiness. This stage transforms the blueprint into reality.

- **Operationalize**: This involves deploying solutions and capabilities into production and officially commencing the data expedition. All members are trained to use the equipment and techniques to scale the summit. The hard work of preparation meets the thrill of execution.

- **Evolve**: Enables continuous improvements through feedback loops and discoveries. Successful climbers observe changing terrain and weather patterns and adapt their approaches dynamically. Similarly, data solutions must evolve based on new insights.

Case Study: Confidential Organizational Transformation

In an anonymized case study, a global retail company embarked on a transformative journey leveraging data and AI to revolutionize its customer experience and operational efficiency. Due to confidentiality agreements, specific details and the organization's name remain undisclosed. However, this case exemplifies the strategic implementation of the data expedition methodology, demonstrating significant achievements in scalability, innovation, and market competitiveness.

- **Discover**: The organization identified underutilized data assets and untapped AI opportunities to enhance personalized customer interactions and streamline supply chain operations.

- **Design**: A collaborative effort between data scientists, engineers, and business strategists led to the blueprinting of sophisticated AI models to predict customer preferences and optimize inventory management.

- **Develop**: The models were refined through iterative development and rigorous validation processes to align with precise business goals, demonstrating promising results in controlled environments.

- **Operationalize**: The transition from pilot projects to full-scale production involved meticulous planning and alignment with existing technological ecosystems, ensuring a seamless integration that facilitated real-time insights and decision-making capabilities.

- **Evolve**: Continuous monitoring and adaptive learning mechanisms were established to maintain the relevance and accuracy of AI applications, driving sustained improvements and value generation.

This case underscores the essential role of a tailored methodology in navigating the complexities of data and AI-driven transformation, securing organizational buy-in, and showcasing measurable outcomes that justify investment in such initiatives.

2. Adapting for Data and AI

Special gear and techniques—like data-focused user stories, flexible experimentation, and MLOps—are necessary to traverse the terrain. **Data and AI projects require specific adaptations of traditional delivery methodologies:**

- **User stories**: Data-focused user stories emphasize data quality, accessibility, and analytics, ensuring data initiatives address real business needs.

- **Experimentation**: Flexible experimentation gathers rapid feedback to improve models, making data and AI projects more agile and responsive.

- **Integration**: MLOps integrates model deployment, monitoring, and management, ensuring that AI models are deployed and managed effectively in production environments.

- **Domains**: Cross-functional teams organized around domains increase alignment and collaboration, empowering teams to tackle data challenges effectively.

With the right equipment and techniques, teams can find the best pathways through the challenging data landscape.

3. Embracing Exploration: From Shu to Ri

The inherently exploratory nature of data initiatives necessitates adaptability in delivery approaches. The Japanese Shu Ha Ri model encapsulates this realization. In the Shu phase, teams strictly adhere to defined methodologies. However, limitations become apparent as data projects progress into the Ha phase. This leads to the crucial Ri phase, where teams transcend rigid practices and embrace flexibility.

For data projects, this means:

- Iterative development based on continuous feedback loops.

- Agile frameworks that allow adaptation.

- Experimentation and prototyping of multiple hypotheses.

- Emphasis on data quality and governance.

- Cross-functional collaboration.

- Proactive risk management.

- Fostering a learning culture across teams.

By fully embracing the Ri phase, teams can navigate uncertainties, unlock insights, and successfully deliver exploratory data initiatives.

4. Aligning with Data Mesh

This approach empowers domain teams while enabling organizational alignment—like experienced climbers balancing autonomy with coordination. **The data mesh approach creates an ecosystem of empowered but aligned teams:**

- **Domain teams**: These own data products that cater to their users, fostering a sense of ownership and accountability.

- **Federated governance**: Grants autonomy while adhering to guidelines, allowing teams to operate independently within defined boundaries.

- **Platform capabilities**: Enables self-service and access to data, promoting collaboration and efficiency across the organization.

Like climbers securely roped together, alignment through data mesh principles prevents teams from straying too far in dangerous directions.

Exploring the data expedition methodology and its application through case studies reveals that certain foundational practices underpin success. These strategic practices serve as the cornerstone for organizations aiming to navigate the complexities of data and AI projects effectively. Let us delve into these practices and understand their impact through the lens of industry leaders and innovative companies.

5. Climbing With Agility

Agility is paramount in the journey of data and AI exploration. Agile workflows are a testament to the industry's shift towards more dynamic, responsive, and user-centric development processes. Agile principles facilitate a smoother expedition through the data landscape and ensure that outcomes are aligned with user needs and business objectives.

Agile workflows like Scrum, retrospectives, and backlog prioritization help teams navigate obstacles and choose optimal routes—just as mountaineers do. Agile events and artifacts boost collaboration, feedback cycles, and continuous delivery.

- **Prioritized backlogs**: Data needs are ranked by value and complexity, enabling teams to focus on high-impact initiatives.

- **Sprints**: Initiatives are broken into achievable milestones, providing a clear path to progress and success.

- **Standups**: Enables alignment, blocker identification, and planning, ensuring teams stay on track and proactively address challenges.

- **Retrospectives**: These collect improvement ideas and learnings, fostering a culture of continuous improvement and collaboration.

With agile workflows, data teams can dynamically adapt and find the fastest path to the summit.

- **Implementation**:

  - **Prioritized backlog**: Begin with a prioritized backlog that ranks data needs by value and complexity.

  - **Sprints**: Create time-boxed sprints to break initiatives into achievable milestones.

  - **Standups**: Hold daily standups to align the team, identify blockers, and plan the next steps.

- **Key metrics**:

  - **Sprint velocity**: This measures the amount of work completed in each sprint, helping to predict the team's future performance. An increasing velocity suggests improved efficiency.

  - **User story completion rate**: This metric focuses on how many user stories the team completes per sprint. A high rate signifies strong alignment with stakeholder needs.

  - **Cycle time**: Measures the time taken from identifying a task to its completion, offering insights into process efficiency.

  - **Escaped defects**: Tracks defects that make it to production, helping to gauge the quality of work.

  - **Business value delivered**: This measures the value of the features delivered, helping to align the development work with business goals.

Case Study: The Agile Revolution and Data-Driven Personalization

Spotify, renowned for revolutionizing music streaming, managed rapid product growth while fostering innovation and swiftly responding to market shifts. Central to addressing these challenges was adopting the agile methodology and an advanced, data-driven approach for personalizing music experiences.

- **Agile transformation**: Spotify introduced the Spotify Model, an agile working framework emphasizing autonomy, rapid adaptation, and seamless communication. **This model is distinguished by:**

  - **Squads**: Cross-functional teams dedicated to specific features.

  - **Tribes**: Collections of squads working on interconnected features.

  - **Chapters and guilds**: Structures promoting collaboration and ongoing learning across different domains.

- **Data-driven personalization**: To tailor music experiences, Spotify employed:

  - **Collaborative filtering**: Analyzing user behaviors to identify patterns and recommend songs.

  - **Content-based filtering**: Making suggestions based on song and artist attributes.

  - **ML algorithms**: Utilizing audio features to discern and anticipate user preferences.

- **Outcomes**: The Spotify Model significantly contributed to a vibrant and innovative work environment, facilitating swift feature development and deployment. The personalized approach profoundly enhanced user engagement, with custom playlists and recommendations ensuring a distinct listening experience for each user.

Spotify's agile transformation and data-centric strategies have solidified its position as a market leader and set a benchmark for digital and product innovation across various sectors (Intelloz Consulting Group 2023; Althris Training 2023; Reqtest 2015; Lopez 2023; Gardner 2023; Al Farabi 2023b).

This case study exemplifies the impact of integrating agile workflows with data and AI on achieving scalable personalization. It highlights Spotify's adaptive and iterative development process in response to user behavior and trends, underscoring the principles of flexibility and adaptability in digital innovation.

6. Putting the User First

In the ascent of any data and AI initiative, the user's perspective stands as the north star, guiding each decision and innovation. Design thinking emerges as a critical methodology for ensuring that technological advancements remain cutting-edge and deeply resonant with those they are meant to serve. It is based on empathy, creativity, and iterative testing, forming the bedrock of user-centric solutions.

Climbers understand that success means getting themselves and their team to the summit. **For data teams, this means profoundly understanding user needs and rapidly iterating on solutions:**

- **Research**: Reveals user needs and pain points, allowing data teams to build solutions that address user requirements.

- **Brainstorming**: Develops creative data solutions, encouraging innovative approaches to data challenges.

- **Rapid prototypes** validate concepts through honest user feedback, ensuring data solutions align with user expectations.

- **Iterative enhancement**: Continuously improve data utility, which allows teams to refine and optimize solutions based on user feedback.

With design thinking, teams build solutions users want and will use to reach the summit.

- **Implementation**:

  - **Interviews**: Conduct user interviews to identify pain points.

  - **Prototype solutions**: A/B tests them for user feedback.

  - **Incorporate feedback**: Refine data solutions based on feedback.

- **Key metrics**:

  - **A user satisfaction score** gauges user opinion of a data product or feature. Higher scores indicate more effective design thinking implementation.

  - **Iteration cycles**: This measures the number of prototype-test iterations before arriving at the eventual solution, giving insights into the project's agility and user-centric focus.

  - **Net promoter score (NPS)**: Measures customer loyalty and satisfaction, providing another perspective on user experience quality.

- **Conversion rates** measure how well the design converts users to desired actions, such as app or platform sign-ups or purchases.

Design thinking aligns with the Ri phase, enabling teams to continuously adapt solutions based on user feedback rather than rigidly following predefined practices.

Case Study: IBM

IBM shifted from focusing primarily on hardware to a diversified portfolio that included software, services, and solutions. This move aimed to foster innovation and enhance customer satisfaction within the competitive technology landscape.

- **Challenge**: To remain relevant in the fast-evolving tech industry, IBM recognized the need to enhance its product usability and customer experience. The organization aimed to transition from a product-centric approach to a user-centric design philosophy.

- **Solution**: By embracing design thinking principles, IBM now focuses on profoundly understanding and empathizing with user needs. This strategy involves forming multidisciplinary teams, including designers, engineers, and business stakeholders, to facilitate diverse perspectives and collaborative problem-solving. IBM's commitment to design thinking also established the IBM Design studios, hubs for applying these principles across product development and customer engagement strategies.

- **Implementation**: IBM applied design thinking to areas such as its enterprise software solutions to focus on solving user problems and improving the user interface for better accessibility and satisfaction. Projects like Watson (IBM's AI platform) showcased the approach's benefits, resulting in technologically advanced, intuitive, and user-friendly solutions.

- **Outcome**: IBM's adoption of design thinking revolutionized its product development process, leading to innovative solutions that more effectively met customer needs. This approach has improved client engagement, streamlined product development, and nurtured a culture of innovation and continuous improvement (Bivins 2014; Bjelland & Wood 2008; Human Synergistics 2024; This is Design Thinking n.d.; UserTesting n.d.; Handa & Vashisht 2017).

IBM's strategic embrace of design thinking has set an industry benchmark for integrating user-centric design into corporate culture and operations. This shift has contributed significantly to the company's sustained growth and innovation, reinforcing IBM's position as a leader in using design to drive business strategy and outcomes.

7. Streamlined Operations

The realm of data and AI is inherently dynamic, necessitating an operational backbone that is both resilient and adaptable. DevOps automation represents this backbone by seamlessly integrating development and operational processes to enhance efficiency, reliability, and speed. Automating workflows and embracing an integration and delivery culture can catapult data initiatives to new heights.

DevOps enables data teams to focus on delivery and introduces automation into traditionally manual processes:

- **CI/CD pipelines**: Enable smooth deployment flows, allowing data teams to activate solutions quickly and reliably.

- **Infrastructure-as-code**: Codifies and automates provisioning, simplifying the setup and configuration of data environments.

- **Monitoring**: Provides real-time system observability, ensuring data teams can proactively identify and address issues.

- **Reliability engineering**: Focuses on resilience, ensuring that data solutions are robust and dependable.

With DevOps, data teams spend less time performing routine maintenance and more time scaling the heights.

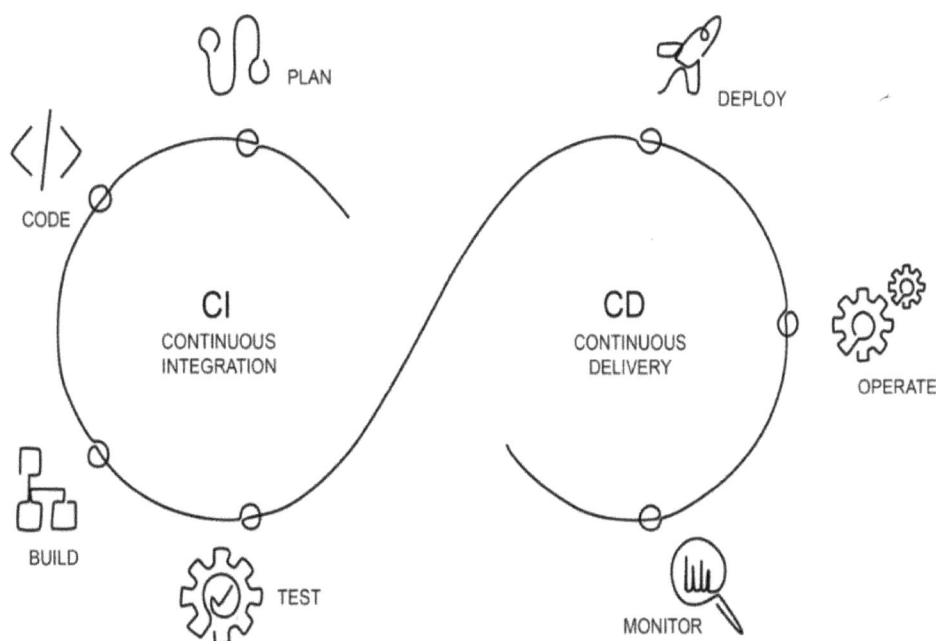

Figure 40. CI/CD Pipeline, Automated Testing, and Monitoring

Case Study: Scalable and Reliable Infrastructure

Netflix's transformation into a DevOps powerhouse traces back to a pivotal event in 2008 when significant database corruption issues underscored the limitations of traditional data center approaches in the face of rapid user base expansion. This catalyzed Netflix's shift towards a more scalable and reliable infrastructure, leading to the adoption of ensuring cloud-native solutions and a microservices architecture.

Emphasizing DevOps principles, Netflix integrated development with operations to ensure swift and dependable service delivery. **Critical practices and tools adopted by Netflix include:**

- **Continuous integration and delivery (CI/CD)**: Automating the software build and test processes ensures that the codebase is always deployment-ready.

- **Infrastructure as code**: Programmatically managing the infrastructure to automate and streamline deployment processes.

- **Containerization**: Simplifying deployment and management by packaging services into containers.

- **Chaos engineering**: Proactively enhancing system resilience by simulating failures such as random service terminations and network outages.

Leveraging these DevOps principles, Netflix achieved the capability to deploy updates multiple times a day, fostering an environment of rapid innovation and continuous improvement. This maintains a robust and reliable streaming platform for a global audience and positions Netflix as an industry leader in utilizing DevOps to drive technological advancement and customer satisfaction (Ensono n.d.; Izrailevsky, Vlaovic, & Meshenberg 2016; Asif 2023; Rakhra 2023; Paul 2020).

In aligning with the Ri phase of the data expedition methodology, Netflix exemplifies how a culture of experimentation and adaptability underpins successful data projects. Their journey illustrates the importance of embracing a flexible approach that allows for scaling, **evolving, and refining operations in the dynamic field of data and AI. Key metrics included:**

- **Deployment frequency**: How often new code is deployed indicates a DevOps pipeline's efficiency and agility.

- **System uptime** is the reliability and availability of services, which is critical for assessing the effectiveness of DevOps operations.

- **Lead time for changes is the time** it takes for a code commit to be deployed into a live environment, which shows process efficiency.

- **Change failure rate**: How often changes fail or require immediate remediation, helping to assess the risk associated with new deployments.

Case Study: DevOps Transformation

Capital One embarked on a transformative journey to modernize its IT infrastructure to address customers' evolving digital banking needs. Recognizing the limitations of traditional banking practices—including slow processes and an inability to keep pace with technological advancements—Capital One adopted a comprehensive DevOps approach.

This shift aimed to enhance operational efficiency, reduce the time to market for new features and improve the reliability of banking services. **Key strategies employed included:**

- **Agile adoption**: Transitioning to agile methodologies to foster flexibility and improved collaboration across development teams.

- **Automation**: Implementing automated testing and deployment workflows to streamline software release processes.

- **Cloud migration**: Capital One moved entirely to AWS to utilize scalable, on-demand infrastructure, making it a pioneer among US banks in cloud migration.

This strategic overhaul has positioned Capital One closer to technology companies than traditional financial institutions. The adoption of DevOps practices enabled Capital One to rapidly develop and deploy banking applications and websites, effectively meeting customer needs. The transformation has fostered a culture of continuous improvement, significantly reducing errors and improving product quality.

Capital One's journey highlights the potential of agile and cloud technologies to overcome the banking sector's operational challenges. By embracing innovation and focusing on continuous improvement, Capital One enhanced its service delivery and established itself as a leader in digital banking (Dhaduk 2022; AWS n.d.-b; Thinklogic n.d.; Kubernetes n.d.).

8. Cloud Computing: The Digital Sherpa

Cloud Computing has revolutionized how organizations approach data storage, processing, and analysis. It offers a scalable, flexible, and cost-effective digital infrastructure. As a digital sherpa, cloud technologies easily guide data and AI projects up the steep inclines of big data and complex computations. The strategic advantages of cloud computing in data initiatives include on-demand resources, managed services, and platform scalability.

The cloud provides data teams with pre-provisioned resources and guided paths to the summit:

- **Adaptation**: On-demand data and compute resources tailored to varying needs, allowing data teams to scale resources as required.

- **Guidance**: Managed data/AI services provide expert guidance without the internal overhead, enabling data teams to leverage best practices.

- **Capacity**: Scalable storage houses vast volumes of data, providing the capacity to handle large datasets.

- **Maximizing ROI**: Cost optimization tools increase efficiency and value, helping organizations maximize the return on their data investments.

The cloud removes heavy burdens so data teams can focus on the final objective. Cloud resources' flexibility supports the Ri phase, allowing teams to scale their data exploration as insights emerge dynamically.

Case Study: Amazon AWS

Amazon's adoption of Amazon Web Services (AWS) marked a significant transformation from traditional on-premises database management to a pioneering cloud-based architecture. This move underlined the global trend toward embracing cloud computing due to its scalability, reliability, and cost efficiency.

- **Challenge**: Initially, Amazon's infrastructure was heavily reliant on Oracle databases, which presented substantial management, provisioning, and capacity planning challenges due to the company's exponential growth. The need for a more scalable and efficient database solution became a priority.

- **Solution**: Amazon initiated a bold move to migrate over 5,000 databases to AWS. This strategic pivot was a technical migration and a fundamental shift towards leveraging AWS's comprehensive cloud capabilities, including services such as Amazon DynamoDB, Amazon Aurora, Amazon S3, and Amazon RDS.

- **Implementation**: The migration process was meticulous, requiring technical adjustments and fostering a cultural shift within Amazon towards cloud-centric skills and methodologies. This transition aimed to capitalize on AWS's scalable infrastructure to achieve a 50 percent reduction in operational costs and a 40 percent improvement in latency for critical services despite increasing transaction volumes.

- **Outcome**: The switch to AWS dramatically decreased database administration overhead and operational expenses, establishing a new standard for scalable, high-performance cloud solutions. Through the integration of AWS services, Amazon enhanced its data management capabilities, ensuring greater availability, durability, and performance.

- **Impact**: Amazon's migration to AWS serves as a testament to the transformative power of cloud computing in modernizing business operations. It exemplifies how strategic digital transformation can significantly enhance scalability, cost efficiency, and operational effectiveness (AWS n.d.-a; Hespell 2023; Kovalskiy 2024; Walton 2016; Sinha 2023; Viégas, Holbrook, & Wattenberg 2018; Patel & Shetty 2024; Chatterjee 2024; Clarke 2024).

- **Key metrics**:

  o **Cost optimization rate**: How resources are utilized relative to their costs is critical for assessing cloud ROI.

  o **Resource utilization**: How effectively computing and storage resources are used, helping with cost control and scaling decisions.

  o **Data throughput**: The volume of data that can pass through a system in a given time indicates performance capabilities.

  o **Latency**: The time it takes for a data packet to move from source to destination is critical for user experience and system performance.

9. Measuring Outcomes

In mountaineering, climbers continually assess conditions and progress. Similarly, metrics and KPIs are the compasses for data/AI projects, guiding teams toward their objectives and providing early warnings when course corrections are needed.

- **Data value metrics**: Demonstrate business impact, showcasing the value of data initiatives to the organization.

- **ROI assessments**: Evaluate costs versus benefits, helping organizations make informed decisions about their data investments.

- **Feedback loops**: Provide user perspectives, enabling data teams to understand how well solutions meet user needs.

- **Maturity benchmarking**: Assesses capabilities and gaps, guiding organizations on their path.

With measurable outcomes, data teams can calibrate their course and prove the expedition's success to stakeholders.

- **Implementation**:

    o **Data value metrics**: Implement these to track business impact.

    o **ROI assessments**: Use these to evaluate costs versus benefits.

    o **Maturity benchmarking**: Utilize these to assess capabilities and gaps.

Metric Category	Metric	Description
Agile Metrics	Sprint Velocity	Measures the work completed in each sprint, providing insights into team efficiency.
	Burndown Chart	Visualizes work remaining, enabling proactive adjustments.
Design Thinking Metrics	User Satisfaction Score	Captures user sentiment, offering insights into how well a solution meets its intended purpose.
	Net Promoter Score (NPS)	Gauges customer loyalty and overall satisfaction with the end product.
DevOps Metrics	Deployment Frequency	Monitors how often new features, fixes, or updates are released.
	Lead Time for Changes	Measures the time from code commit to code successfully running in production.
Cloud Computing Metrics	Cost Optimization Rate	Measures how healthy resources are utilized relative to costs.
	Resource Utilization	Monitors cloud services and resources, aiding in effective scaling decisions.
Overall Data/AI Project Metrics	ROI (Return on Investment)	Captures the financial benefits of the data/AI projects.
	Model Accuracy	Predictive or classification accuracy is a critical metric for AI models.
	Data Freshness	Ensures data is up-to-date and relevant for ongoing analytics.

Table. Critical Metrics for Navigating Success in Agile, Design Thinking, DevOps, and Cloud Computing Projects

Case Study: Google's AI Impact on Business and Society

Google's integration of AI across its offerings showcases a pioneering commitment to enhancing user experiences, streamlining operations, and extending societal benefits through technology.

- **Challenge**: Determining the direct ROI from AI initiatives is challenging, as their impact often extends beyond financial metrics to broader engagement, efficiency, and societal advancements.

- **Solution**: Google has woven AI into its core structure, infusing these technologies into search algorithms, advertising, cloud services, and societal advancement initiatives. These applications aim to refine products and services and tackle global challenges through innovative solutions.

- **Implementation**: Google's AI strategies span various sectors:

  o **Healthcare**: AI-driven research and tools are revolutionizing diagnostics and treatments.

  o **Environmental sustainability**: AI assists in monitoring ecological changes, aiding conservation efforts.

  o **Accessibility**: AI-powered tools improve access to information and technology for those with disabilities.

- **Outcome**: The initiatives have led to advancements in technological applications, enhanced user experiences, and notable societal contributions. While direct financial impacts are nuanced, the broader implications underscore AI's capacity to deliver significant value across multiple dimensions (Shahid, 2023; Kovalskiy, 2024; Walton, 2016; Sinha, 2023; Viégas, Holbrook, Wattenberg, 2018; Patel & Shetty, 2024; Google Cloud n.d.-b).

Google's deployment of AI reflects a nuanced approach to valuing technological implications. It stresses the importance of aligning AI projects with business objectives and societal needs to foster innovation and transformative benefits.

Conclusion

By embracing the methodologies and insights of this chapter, organizations will be well-prepared to scale the heights of data and AI excellence. The data expedition methodology provides a structured approach that promotes scalability, aligns with business objectives, and ensures iterative development and refinement of data solutions. **Key takeaways include:**

- **Structured progression**: The methodology emphasizes a phased approach, from discovery to deployment and continuous improvement, ensuring that data projects are strategically planned and executed.

- **Adaptability and flexibility**: Incorporating agile workflows and iterative development allows organizations to remain responsive to changing data landscapes and business needs.

- **User-centric design**: Prioritizing user needs and incorporating feedback throughout the development process ensures that the solutions are practical and widely adopted.

- **Operational efficiency**: Leveraging cloud computing and DevOps practices enhances operational efficiency, enabling rapid deployment and the reliable performance of data solutions.

- **Governance and responsibility**: Implementing robust governance practices ensures that data and AI initiatives are conducted with integrity and accountability and address ethical considerations and compliance requirements.

Anchoring these endeavors within a robust governance framework becomes crucial. This framework serves as a compass that can be depended upon while traversing the complex landscape of data and AI. It ensures that exploration is conducted with integrity and responsibility towards society.

The lessons of the data expedition methodology must be carried forward into governance practices to ensure that the innovation journey is both successful and responsible.

CHAPTER 13

Staying Safe on the Journey

Without the ropes, harnesses, and other safety gear that professional climbers rely on, ascending a mountain would be an extreme gamble. Moving higher up the peak, they will encounter treacherous terrain, unpredictable weather, and daunting challenges.

Likewise, organizations engaged in data and AI initiatives require responsible AI governance practices to provide the guardrails to prevent catastrophic falls. Beyond safety, data ethics and accountable AI also illuminate the path toward socially conscious and trustworthy decision-making. Data and AI governance are the keys to staying safe and thriving on this thrilling ascent (Engler 2022; Meier & Spichiger 2024; European Parliament 2024).

Fairness, transparency, and accountability resonate deeply with the spiritual values of integrity, compassion, and harmony. Our technological endeavors should advance our capabilities, enhance our collective well-being, and connect us more deeply to one another and the planet.

Intertwining technology with our deepest values will guide us toward decisions that uplift humanity and foster a profound sense of unity and ethical stewardship. This is not just about conquering technological challenges but about realizing the role of data and AI in reflecting our collective aspirations for a world grounded in consciousness and spirituality. Viewing data governance through the lens of stewardship echoes spiritual traditions in which stewardship is a sacred trust. Data is not merely an asset but a collective heritage that must be managed wisely and ethically, reflecting a commitment to community welfare and the integrity of our planet.

The potential of AI to either erode or enhance human connections is profound. Mindfully designed, AI can bridge cultural divides and foster empathy, creating a tapestry of human experiences that enrich our spiritual journey. Projects that leverage AI for social and environmental betterment give glimpses of the potential of technology to unite us in the shared human quest for understanding and compassion.

The Consequences of Poor Governance

The risks of neglecting data and AI governance can be catastrophic. Consider a company that experiences a massive data breach that exposes sensitive customer information to malicious actors. The aftermath is a frantic scramble to contain the damage, regain customer trust, and grapple with regulatory fines. Without proper data governance, this scenario becomes a reality.

When climbing a mountain, there are moments of uncertainty when the ground beneath your feet feels unsteady, and every step becomes a high-stakes gamble. Organizations venturing into uncertain data and AI terrain are likewise exposed to potential pitfalls and hazards.

The creation and deployment of AI systems call for mindfulness—a conscious consideration of what technology can and should do for the greater good. This mindful approach fosters technology that supports personal growth and spiritual awakening, ensuring that AI catalyzes connection rather than division.

While governance provides the overarching policies and guidelines, metadata management and MLOps provide crucial technical capabilities that enable organizations to operationalize responsible data and AI systems effectively. Metadata is the informational scaffolding that makes data understandable and traceable, providing critical lineage and context to support governance. MLOps pipelines allow organizations to deploy and manage AI models in alignment with governance protocols around risk, ethics, and continuous monitoring requirements. Governance, metadata, and MLOps work synergistically to establish accountability while enabling technical teams to responsibly build and operate data products.

Metadata management and MLOps pipelines are crucial components in the lifecycle of AI and data analytics projects. They work synergistically to enhance accountability, improve efficiency, and use data and ML models responsibly. Below are examples of how both can be implemented to work together.

Metadata Management Examples

Data Cataloging

A data catalog provides a centralized repository for organizing, accessing, and managing metadata across an organization. It allows data scientists and analysts to quickly find relevant data, understand its source, quality, and lineage, and know who is responsible.

- **Example**: Collibra ensures data trust and compliance, offering centralized repository functionalities for organizing, accessing, and managing metadata, thus fostering adherence to governance standards (Weldon 2024).

Data Lineage Tracking

Data lineage tracks data flow from its source through various transformations to its final form. This is essential for debugging, auditing, and understanding the impact of changes in data pipelines.

- **Example:** Informatica and Talend provide data lineage tracking capabilities, allowing organizations to trace data through various transformations, which is crucial for maintaining data integrity and supporting regulatory compliance efforts.

MLOps Pipeline Examples

Continuous Integration and Deployment (CI/CD) for ML Models

In the context of ML, CI/CD involves automatically testing, validating, and deploying ML models to production environments. This ensures models are reliably updated without disrupting the application they are powering.

- **Example**: Jenkins and GitLab CI, when used with MLflow, facilitate the CI/CD pipeline for ML models, automating testing, validation, and deployment processes to maintain models' effectiveness in production.

Model Monitoring and Versioning

Continuous monitoring of deployed models is necessary to catch performance degradation over time, known as model drift, and to ensure that the model's predictions remain valid as underlying data changes.

- **Example**: Prometheus and Grafana offer monitoring solutions for ML models, while DVC provides version control for datasets and ML models, ensuring robust model management and operational efficiency.

Consider the transformative power of AI when aligned with spiritual values. Companies can prioritize ethical AI to foster workplace harmony, initiatives to leverage technology for ecological restoration, or platforms that use AI to connect individuals across the globe in meditation and mindfulness practices. These examples showcase AI's potential for good and its role in advancing our spiritual evolution.

Synergistic Operations

In essence, metadata management provides the necessary context and understanding of data, ensuring it is used correctly and responsibly. In contrast, MLOps pipelines offer the framework and processes for practically applying this data in ML models. They form a robust foundation for accountable, **efficient, and compliant AI development and deployment via several avenues:**

- **The feedback loop between metadata management and MLOps:** Metadata from data management systems, like data quality scores and lineage, can feed into MLOps processes, informing model retraining decisions. For example, if data lineage reveals that a data source has changed significantly, an MLOps pipeline can automatically trigger the retraining of the associated models.

- **Governance and compliance:** Metadata management ensures that data used in training and inference phases is documented, traced, and complies with privacy regulations (like GDPR). Simultaneously, MLOps pipelines ensure that models are developed, deployed, and operated according to these data governance rules. This dual approach reinforces accountability and compliance across all stages of the AI model lifecycle.

- **Enhanced collaboration and efficiency:** Both systems promote collaboration among data scientists, ML engineers, and governance teams by centralizing metadata and streamlining ML workflows. Metadata management clearly understands data, while MLOps ensures that models are efficiently deployed and maintained. This collaboration minimizes errors, reduces development cycles, and ensures responsible usage and operation of data products.

Examples of High-Profile Data Breaches

Here are some examples of high-profile data breaches that demonstrate the real-world consequences of poor data governance:

- **Facebook-Cambridge Analytica scandal (2018):** The political consulting firm Cambridge Analytica improperly accessed and misused the data of up to 87 million Facebook users. This massive breach of privacy was enabled by lax governance over app developer access to user data. This egregious privacy breach was primarily enabled by Facebook's lack of stringent governance over its extensive data repositories and third-party access controls, allowing Cambridge Analytica to amass detailed psychographic profiles on millions without explicit consent. The consequences for Facebook were severe: the company faced billions of dollars in market cap decline, significant fines, widespread public outrage, and irreparable damage to its reputation (Meredith 2018).

- **Equifax breach (2017)**: The sensitive personal information of approximately 143 million people was exposed when Equifax failed to patch a known security vulnerability in its systems (Fruhlinger 2020; Lindsey 2019).

- **Marriott International (2018)**: The data of up to 500 million hotel guests was accessed in a breach attributed to insufficient data governance controls around customer records inherited in an acquisition (Schneider 2018; Faife 2022).

- **British Airways (2018)**: The airline suffered a data breach 2018 that affected 380,000 to 500,000 customers. Hackers accessed customers' personal and financial details and made bookings (Burgess 2018).

In contrast, recent successes in responsible AI applications—such as AI in healthcare for predicting patient outcomes while ensuring data privacy—demonstrate how ethical considerations can lead to innovative solutions that respect individual rights and societal values.

Recent Developments in Responsible AI

Recent advancements in AI, including sophisticated models like GPT-4, have brought significant challenges related to privacy, bias, and data security to the forefront, necessitating updated governance strategies to navigate the complex landscape of AI ethics and legal compliance.

The introduction of models capable of generating highly realistic and nuanced content has raised concerns about the potential for privacy invasions, as AI can produce detailed profiles of individuals without explicit consent. Furthermore, the risk of perpetuating or exacerbating societal biases through AI algorithms has become a prominent issue. Models trained on historical data may inadvertently learn and replicate biases present in that data, leading to unfair outcomes in applications ranging from hiring practices to judicial sentencing.

Moreover, the security of data used in training and operating these AI systems is paramount. As models become more integrated into critical infrastructure and personal applications, the potential impact of data breaches and misuse grows. Ensuring the security of systems against unauthorized access and manipulation is a crucial aspect of responsible AI development and deployment.

To address these challenges, technologists, ethicists, policymakers, and the public are engaged in an ongoing dialogue about the best paths forward. The European Union's Artificial Intelligence Act is a pioneering effort to set comprehensive standards for AI development and its use within the EU. The Act categorizes AI systems based on risk levels and sets out specific requirements for high-risk AI applications, emphasizing transparency, accountability, and data protection.

The discourse around AI advancements also emphasizes the need for ethical frameworks and guidelines to shape AI research and applications. Organizations such as IEEE, the European Union, the Partnership on AI, and OpenAI have developed principles and standards to foster AI systems' ethical development, deployment, and governance.

The complex interplay of AI advancements and societal values requires a multifaceted approach involving collaboration across sectors and disciplines. By fostering open dialogue, implementing rigorous governance frameworks, and prioritizing AI's ethical and secure use, AI's transformative potential can be harnessed while safeguarding individual rights and societal well-being (Meier & Spichiger, 2024).

Federated Governance in the Context of Data Mesh

Different roles—lead climber, belayer, navigator—are clearly defined in a mountain climbing expedition. Likewise, federated governance assigns distinct responsibilities to various stakeholders to form a coordinated approach. In a data-centric organization, managing data assets effectively and ethically is crucial. In the data mesh framework, a federated governance model provides a decentralized approach that allows various units within an organization to own and govern their data while still adhering to centralized policies and guidelines. Data stewardship distributes responsibilities across different organizational domains or departments rather than a single, centralized data team responsible for all governance tasks. Each unit or domain owns its data assets and is responsible for quality, security, and ethical considerations.

While data collection is essential for AI models, it raises ethical questions. For instance, how far can one go in collecting personal information before it becomes an invasion of privacy? Companies must navigate these murky waters carefully. Biases in training data can inadvertently lead to AI models that reinforce existing societal biases, such as racial or gender bias. How do we prevent the perpetuation of such biases in AI systems?

The idea of AI systems making life-or-death decisions in applications like self-driving cars or medical diagnoses is a topic of ongoing debate. The ethical ramifications are immense and warrant careful consideration.

- **Why federated governance?**
  - **Scalability**: As organizations grow, so do the volume and complexity of data. A federated model allows for scalability.
  - **Responsiveness**: With data owners closer to the operational aspects, decision-making is quicker and more responsive to immediate needs.
  - **Ethical and legal compliance**: Each unit is responsible for its data and meeting all moral and legal obligations.
- **How federated governance works**:
  - **Data ownership**: Each domain or department owns its data and is responsible for its quality and usage.

○ **Central policies**: A set of global policies and standards are maintained centrally to which all domains must adhere.

○ **Local implementation**: Local data stewards ensure compliance with global policies but can also implement domain-specific ones.

○ **Audit and compliance**: Regular audits ensure that all units adhere to global and local policies.

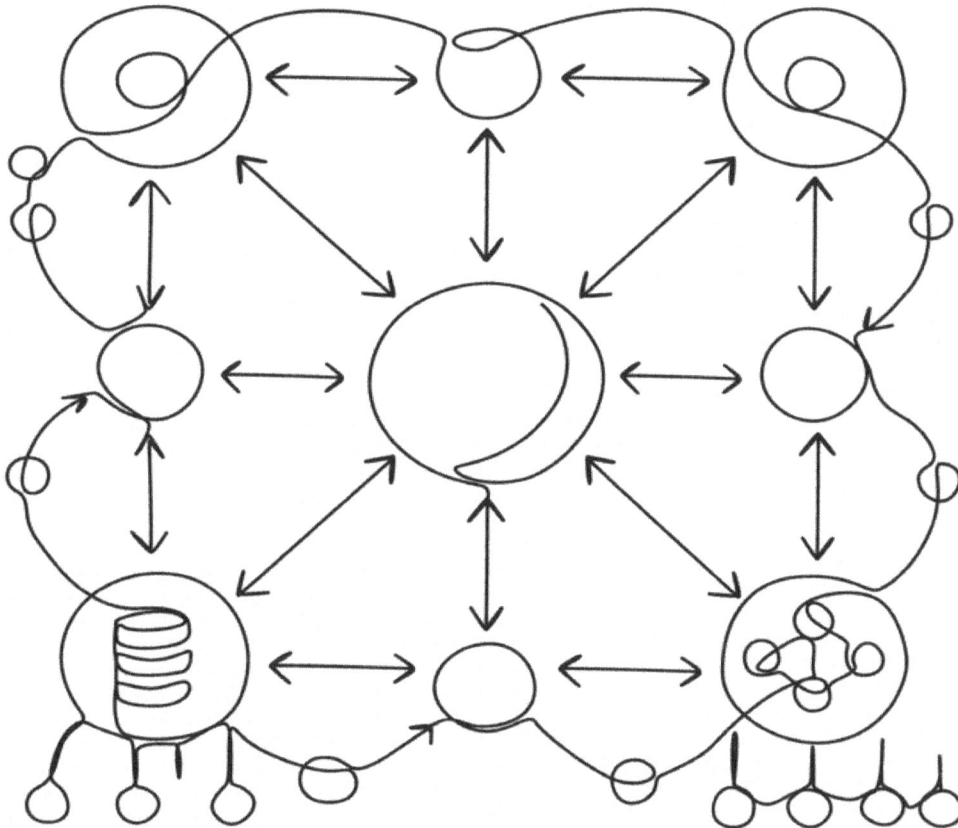

Figure 41. Federated Governance and Data Mesh

Key Components of Data/AI Governance

- **Choosing the right peak to ascend**: Just as climbers select a mountain based on their goals and capabilities, organizations must align their data strategy with business objectives. This data strategy alignment is further enriched by adopting principles of federated governance from data mesh, empowering domain teams to own and manage their data products while adhering to overarching governance guidelines.

- **Establishing base camp rules**: Like climbers establishing base camp rules to ensure safety and coordination, data governance sets policies and standards for data quality, security, and model-risk frameworks. When integrated with federated governance principles, these rules provide a solid foundation for decision-making and ensuring data is trustworthy and reliable across the organization.

- **Designating sherpas for guidance**: On a climbing expedition, experienced sherpas guide and support climbers, ensuring a safe and successful journey. Likewise, federated governance principles designate roles and responsibilities like data owners, data stewards, and privacy/ethics leads. These functions oversee data-related activities within their domains while aligning with broader governance objectives.

Role	Responsibilities	Key Skills/Attributes
Data Governance Board	Overall strategy, policy approval, budget allocation	Strategic thinking, Decision-making, Leadership
Chief Data Officer (CDO)	Implementation of governance strategy, stakeholder liaison, alignment of business and data needs	Leadership, Data expertise, Communication
Data Steward	Ensuring data quality, metadata management, data cataloging	Attention to detail, Data management, Domain expertise
Data Analyst	Data preparation, analysis, reporting	Analytical skills, Data manipulation, Communication
Data Scientist	Advanced analytics, model development	Analytical skills, Programming, Data modeling
Data Ethics Committee	Evaluation of ethical implications, oversight of ethical guidelines	Ethics expertise, Critical thinking, Compliance
AI Operations Team	Deployment and monitoring of AI models, ensuring performance and fairness	Technical skills, AI expertise, Monitoring

Table: Summary of Governance Roles & Responsibilities

- **Adhering to climbing regulations**: Organizations must meet regulatory compliance requirements and conform to data privacy laws and industry regulations. GDPR, CCPA, and other data protection laws are the ethical compass for data and AI initiatives. Federated governance ensures compliance at the domain level.

- **Navigational tools for the journey**: Metadata management is the navigational tool for data initiatives. Data dictionaries, lineage maps, and metadata repositories provide critical insights into the origin, quality, and usage of data, helping organizations make informed decisions. Within a federated governance framework, domain-specific metadata further aids in understanding and managing data within its context.

- **Ensuring equipment is in good condition**: Climbers check their equipment for safety and functionality before ascending. Similarly, data quality practices ensure accurate, consistent, and relevant data. Quality data is the bedrock for trustworthy analytics and AI models, a principle amplified by federated governance through domain-specific data validation and quality monitoring.

AI in Data Quality and Governance

In an era of ever-expanding data, more than manual efforts are required to keep up. AI can automate, enhance, and redefine data quality and governance efforts.

- **Automating data quality checks**:

  - **Outlier detection**: AI algorithms can automatically identify outliers or anomalies in data sets, flagging them for further inspection. This is particularly useful for financial or transactional data where anomalies could indicate fraud.

  - **Data cleansing**: ML models can be trained to identify and correct errors in data records, such as misspellings, duplicates, or missing values, thereby streamlining the data cleansing process.

  - **Semantic understanding**: Natural language processing can be used to understand the context and semantics of text-based data, improving the quality of categorical or textual data points.

- **Enhancing data governance**:

  - **Metadata management**: AI can auto-generate metadata tags and manage metadata libraries, making cataloging and searching easier.

  - **Access control**: ML models can predict which employees will likely need access to specific data types, automating access control decisions while adhering to data privacy regulations.

  - **Compliance monitoring**: AI algorithms can continuously monitor data transactions to ensure they align with regulatory requirements, providing real-time alerts in case of non-compliance.

- **Practical tips**:

  - **Pilot testing**: Start with a small-scale pilot project to test out AI-driven data quality and governance initiatives.

  - **Staff training**: Ensure teams are adequately trained to interpret the AI's findings and intervene when human expertise is needed.

o **Ongoing monitoring**: Continually update AI models to adapt to evolving data patterns and regulatory requirements.

- Identified areas where AI can enhance data quality and governance.
- Conducted pilot testing of AI-driven quality checks and governance policies.
- Staff trained in AI tools and their implications.
- Established a system for the ongoing monitoring and updating of AI tools.

Figure 42. Checklist for AI in Data Governance

By leveraging AI in data quality and governance processes, operations can be streamlined, and valuable human resources can be freed up for more complex, decision-centric tasks. The result is a more agile, compliant, and data-empowered organization.

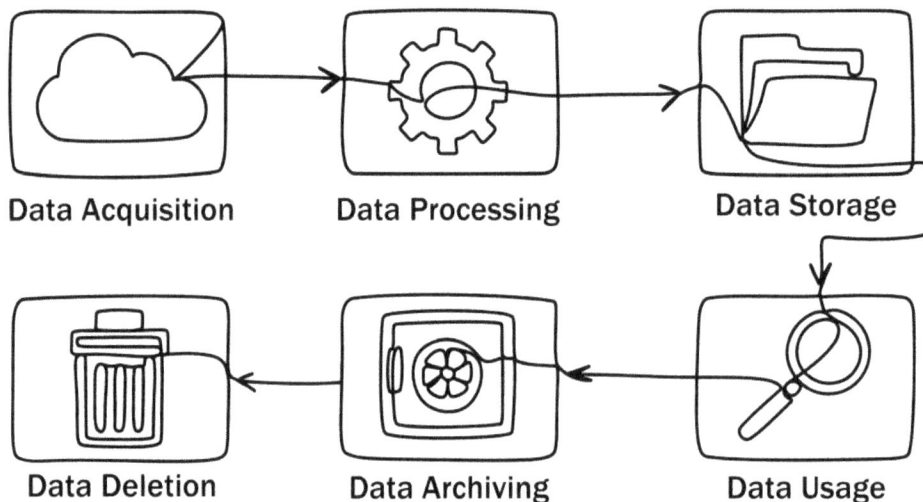

Data Acquisition   Data Processing   Data Storage

Data Deletion   Data Archiving   Data Usage

Figure 43. Data Lifecycle Management

Implementing Governance: Policies and Guidelines in Action

To prevent catastrophic failures in data and AI initiatives, organizations must implement robust governance policies and guidelines that cover a wide range of areas, from data handling and privacy to AI ethics and model management. Here are several examples of such policies and guidelines, organized by key areas of focus:

- **Data privacy and security**:

  o **Data encryption policy**: Mandate sensitive data encryption— at rest and in transit—to protect against unauthorized access.

- **Access control policy**: Using role-based access control (RBAC) systems, define who can access data, under what conditions, and with what level of permissions.

- **Data anonymization guidelines**: Establish procedures for anonymizing personal data, particularly in datasets used for training AI models, to comply with privacy regulations like GDPR or CCPA.

- **Data quality**:

  - **Data validation rules**: Implement automated validation rules to ensure the accuracy, completeness, and consistency of any data entering the system.

  - **Data cleaning procedures**: Review and clean data regularly to remove duplicates, correct errors, and update outdated information to ensure AI models are trained on high-quality data.

- **AI ethics and fairness**:

  - **Bias detection and mitigation framework**: Develop a framework for assessing AI models for bias and implementing mitigation strategies to ensure fairness in AI outcomes.

  - **Transparency guidelines**: Require documentation of AI model development processes, including data sources, model choices, and decision-making criteria, to ensure transparency and accountability.

  - **Ethical AI review board**: Establish an interdisciplinary board responsible for reviewing AI projects for moral considerations, potential societal impacts, and compliance with internal and external ethical guidelines.

- **Model governance**:

  - **Model lifecycle management policy**: Define processes for the entire lifecycle of AI models, from development and testing to deployment, monitoring, and retirement, ensuring models remain accurate and relevant.

  - **Continuous monitoring procedures**: Identify and address model drift or performance degradation quickly.

  - **Version control guidelines**: Ensure reproducibility, allow rollback to previous iterations and track changes over time.

- **Regulatory compliance**:

  - **Compliance checklist**: Create a checklist of relevant regulations (e.g., GDPR, HIPAA) and ensure all data and AI practices comply with legal requirements.

  - **Data protection impact assessments (DPIA)**: Conduct these for new projects involving personal data to identify and mitigate data protection risks.

- **Training, awareness, and preparation**:

  o **Data literacy programs**: Offer training to improve data literacy across the organization, ensuring all employees understand the importance of data governance and their role.

  o **AI ethics**: Provide training for technical teams to foster an organizational culture prioritizing ethical AI development.

  o **Data breach response plan**: Develop and regularly update a comprehensive protocol for responding to data breaches, including notification procedures, steps to contain and assess the breach, and measures to prevent future incidents.

Implementing such governance policies and guidelines helps organizations build a solid foundation for responsible data and AI initiatives, minimize risks, and ensure that their use of AI technologies aligns with ethical standards, legal requirements, and societal expectations.

This fortifies an organization's data and AI ecosystems against risks and aligns them closely with the principles of responsible AI and federated governance from data mesh. Together, they ensure a holistic approach to ethical, secure, and efficient AI and data usage, setting a solid foundation for innovation and societal benefit.

Protocols for Mitigating Perils

In climbing, risk management is essential to avoid hazards and unexpected challenges. Likewise, data governance incorporates risk management protocols to identify and mitigate potential dangers, including bias, security threats, and ethical concerns. A federated governance approach allows domain teams to tailor risk management strategies to their specific data products. Ethical considerations in AI are crucial in guiding responsible development and deployment.

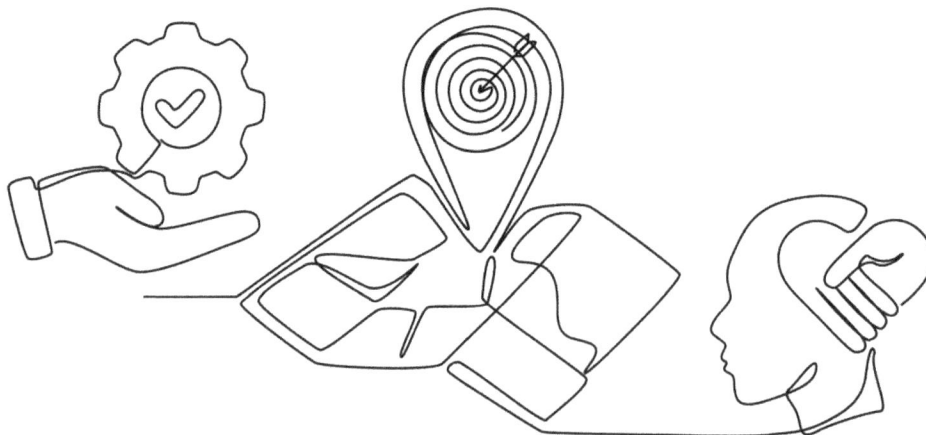

Figure 44. Upholding Integrity and Responsibility on the Digital Frontier

- **Ethical principles in AI**:

  o **Transparency**: Making the decision-making process of AI models understandable to humans.

- o **Accountability**: Assigning responsibility for decisions made by or with the assistance of AI.

- o **Fairness**: Ensuring that AI does not discriminate against any particular group.

- o **Privacy**: Respecting individuals' right to control their personal information.

- o **Safety**: Guaranteeing that AI systems do their intended job without causing harm to humans or the environment.

- **Common ethical dilemmas in AI**:

  - o **Data privacy and consent**: How do we collect and use data ethically?

  - o **Algorithmic bias**: What steps should be taken to identify and mitigate bias in AI systems?

  - o **Job displacement**: How to responsibly manage the human workforce in an increasingly automated environment?

  - o **Environmental impact**: What is the ecological footprint of AI, and how can it be reduced?

  - o **Global inequities**: How to ensure that the benefits of AI are distributed relatively across different communities and nations?

- **Example cases on ethical AI**:

  - o **Healthcare**:

    - ✓ **Problem**: The use of AI in diagnosing diseases.

    - ✓ **Ethical Concerns**: Accuracy, fairness, and data privacy.

    - ✓ **Solution**: In one of my former anonymized projects, a healthcare institution implemented robust validation protocols to ensure the accuracy of AI diagnoses. This involved extensive testing of AI systems with diverse datasets to minimize bias and ensure fairness. Additionally, data privacy was maintained by anonymizing patient data and ensuring compliance with regulations like HIPAA. Regular audits and transparency reports were also produced to inform stakeholders about the AI's performance and ethical considerations.

  - o **Criminal justice**:

    - ✓ **Problem**: Use of AI in predictive policing.

    - ✓ **Ethical Concerns**: Racial bias, transparency, and accountability.

    - ✓ **Solution**: In a project working with an anonymized police department, ethical concerns regarding AI in predictive policing were addressed by several key actions. First, efforts were made to reduce racial bias by incorporating broader, more representative datasets into the AI training process. Additionally, the development team collaborated with external ethicists and legal experts to review the system for potential biases. Transparency was ensured by publishing the criteria and methodologies behind the AI's decisions, enabling public and expert scrutiny. Finally, accountability was enforced by creating an independent oversight body with the authority to audit and, if necessary, suspend AI use if ethical breaches were identified. This process ensured that the AI system adhered to fairness and justice principles.

- **Industry guidelines and standards for ethical AI**:

  - IEEE's Ethically Aligned Design

  - EU Guidelines for Trustworthy AI

  - Partnership on AI's Tenets

  - OpenAI's Charter

- **Strategies for implementing ethical AI**:

  - **Problem**: Use of AI in predictive policing.

  - **Ethical concerns**: Racial bias, transparency, and accountability.

  - **Solution**: In another anonymized project involving a police department, measures were taken to mitigate racial bias by training AI models on diverse and representative datasets. The development team worked closely with ethicists and community representatives to identify and address potential biases. Transparency was ensured by making the AI's decision-making processes and criteria publicly available, which allowed for independent scrutiny. Accountability measures included setting up oversight committees to review the use and outcomes of AI, ensuring that any adverse effects were promptly addressed. These committees also had the authority to halt AI use if ethical standards were compromised.

Ethics in AI are complex but indispensable, and every organization should navigate them with care. By proactively addressing them, companies mitigate risks and pave the way for more responsible and practical AI applications.

Responsible AI: Guiding Stars on the Journey

Data ethics and responsible AI illuminate the path toward socially conscious and trustworthy AI-driven decisions and ensure that the impact on individuals and society is always considered.

Figure 45. Responsible AI: Ethical Decision-Making for AI Systems

- **Responsible AI**: Infusing ethical decision-making into AI development and deployment to ensure transparency, fairness, and accountability. It is a series of practices guiding organizations in creating AI systems that align with human values, legal requirements, and societal norms.

- **Explainable AI**: Since AI models often operate as black boxes, making it challenging to understand their decision-making processes, responsible AI requires explainable AI that provides clear and understandable explanations for outputs. Such transparency helps build trust and allows for proper oversight of AI systems, ensuring ethical decision-making.

- **Bias mitigation**: Responsible AI demands thorough analysis and mitigation of biases to ensure equitable treatment for all. By proactively identifying and addressing biases, organizations can create AI systems that respect diversity and uphold ethical standards.

- **Privacy protection**: Upholding privacy by safeguarding individuals' data. Organizations must adhere to data privacy regulations and ensure that AI models do not compromise people's data security. By establishing strict privacy protocols, organizations can minimize the risks of data breaches and protect sensitive information.

- **Human-centric AI design**: AI solutions should be designed with humans in mind. Responsible AI prioritizes users' experience, needs, and feedback during development. By involving users in the design process and continuously seeking their input, organizations can create AI solutions that genuinely address real-world problems and add value to people's lives.

- **Ethical decision frameworks**: Principles and values frameworks provide a systematic approach to aligning AI systems with ethical considerations and societal norms, ensuring that AI operates responsibly in various contexts.

To further embed responsible AI into the organizational fabric, establish forums and workshops for dialogue on AI ethics, inviting participation from across societal and academic spheres. This encourages a culture of continuous ethical reflection and adaptation.

Responsible AI in Action

The data expedition methodology consists of several stages that outline how data ethics and responsible AI are secured, the roles and responsibilities of all involved, and how such a framework is embedded into the organizational design.

Discover Stage: Laying the Ethical Foundation

Establishing the ethical foundation of the data initiative in the discovery stage is crucial. This involves identifying potential ethical implications and risks associated with the collected data and AI models. Data ethics committees or councils should be formed to review and assess the moral aspects of the project and include subject matter experts, data scientists, ethicists, and legal representatives.

A data ethics committee will conduct ethics impact assessments, ensure compliance with guidelines and regulations, and recommend necessary adjustments to align the initiative with responsible AI practices. In this stage, cross-functional teams should also collaborate to define the AI journey's ethical principles.

Design Stage: Ethical By Design

During the design stage, the data ethics committee plays a pivotal role in ensuring that ethical concerns are embedded in the project from inception. Ethical-by-design principles are applied to stress fairness, transparency, and accountability as essential design requirements.

Data governance frameworks, including policies and standards for data quality, privacy protection, and data sharing, are established. A data stewardship team within the federated governance framework ensures that data is managed ethically and access is controlled to prevent unauthorized use or potential biases.

Develop Stage: Prototyping Responsible AI

As AI models and algorithms are developed, responsible AI practices are applied through iterative prototyping. Explainable AI techniques are incorporated to ensure the transparency of model decision-making, allowing for deeper insights into how models produce their outputs.

The data ethics committee and the AI development team closely monitor potential biases and ethical issues that may arise during model training. This requires ongoing data quality checks and bias mitigation measures. Responsible AI frameworks should be integrated into the AI development process to foster a culture of ethical decision-making among data scientists and AI engineers.

Operationalize Stage: Safeguarding Ethical AI Deployment

During this stage, data stewardship and AI operations teams work in tandem to ensure that models are deployed safely and responsibly. The data stewardship team collaborates with operations to establish monitoring and auditing procedures to continuously assess model performance, fairness, and compliance with ethical guidelines.

For a real-world example, consider healthcare provider XYZ. It utilized ML models to predict patient readmissions. The data stewardship team worked closely with the operations team to ensure fairness in predictions across different demographic groups. This collaboration prevented potential biases and upheld the ethical guidelines the data ethics committee established.

This stage establishes a responsible AI feedback loop, enabling users and stakeholders to provide feedback on AI system outputs. This information is crucial for improving AI models and addressing ethical concerns arising from real-world usage.

Evolve Stage: Ethical AI in Continuous Improvement

Reaching the summit is not the end of a climbing journey. The evolve stage focuses on what comes after reaching initial AI milestones—continuous improvement and adaptation. The evolve stage focuses on constant improvement and responsible AI practices being embedded into the organizational culture. The data ethics committee remains active, continually evaluating the ethical implications of ongoing AI usage and guiding the organization in adapting to evolving ethical standards and regulations.

As new insights and ethical challenges emerge, the organization leverages a learning culture, encouraging open discussions about responsible AI and its potential impact on individuals and society. Responsible AI training and education are provided to employees to foster awareness and understanding of ethical considerations in AI development and usage.

ABC Technologies serves as an illustrative case point. They have established a robust ethical AI culture, continually revisiting their AI ethics guidelines. The tech giant has made responsible AI training mandatory for all its technical staff, embedding ethical considerations in its organizational fabric.

Organizational Design and Responsible AI

Successful implementation of data ethics and responsible AI requires a thoughtful organizational design that embeds these principles into every aspect of the data and AI journey. **Key components include:**

- **Data ethics committees**: Cross-functional with representatives from various departments to ensure ethical considerations are integrated into decision-making processes.

- **Responsible AI champions**: Designating individuals or teams responsible for promoting responsible AI practices and providing guidance on ethical decision-making.

- **Training and education**: Providing comprehensive training and education on responsible AI and data ethics to all employees involved in data and AI initiatives.

- **Ethical AI frameworks**: Developing and implementing frameworks that outline the moral principles and guidelines for AI development and usage.

- **Ethics impact assessments**: Conduct ongoing assessments for all data and AI projects to identify potential ethical risks and develop appropriate mitigation strategies.

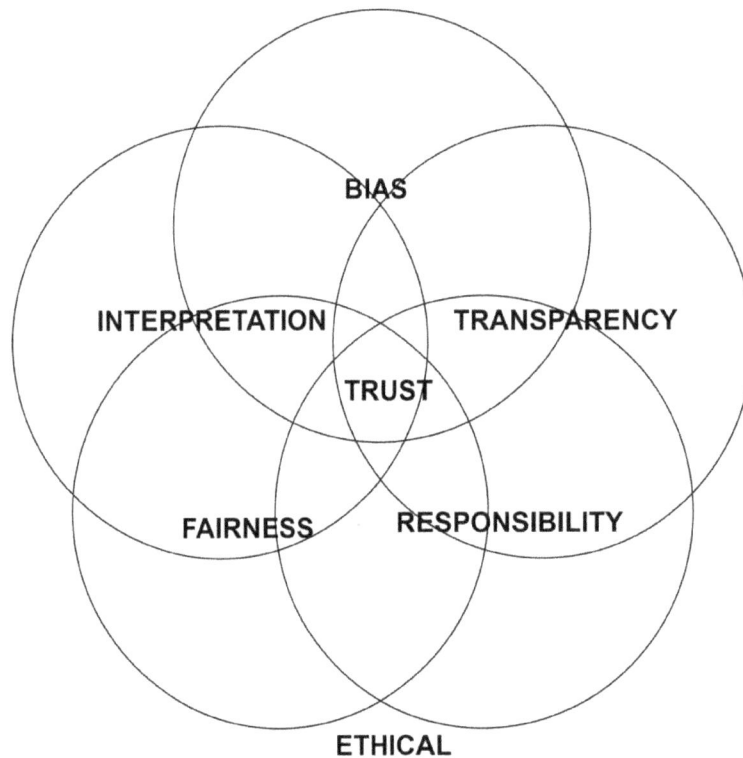

Figure 46. Ethical AI Framework

Incorporating these practical frameworks, methodologies, and organizational designs allows organizations to navigate the data and AI landscape confidently. This approach mitigates risks and paves the way for data and AI to drive positive and transformative impacts on businesses and society.

Conclusion

The frameworks described in this chapter are the safety equipment and guidance mechanisms organizations need to climb the data landscape. To navigate the future, I propose a guiding framework for AI development that embraces spiritual principles. These include holistic well-being, ethical integrity, and the pursuit of wisdom. Organizations should assess their AI initiatives' spiritual and moral dimensions to ensure that technology advances physical capabilities and nurtures our collective needs.

Embracing governance and responsible AI ensures that data is used ethically, securely, and responsibly, leading to positive outcomes for organizations and society. I envision a path beyond mere technological landscapes that touches the zenith of human consciousness and spirituality. The commitment to governance, ethics, and responsible AI is rooted in human values and leads to a future where technology amplifies our capabilities and enriches our sacred essence. Data and AI emerge not as just tools but as companions in our collective quest for a compassionate, connected world.

The ethical considerations discussed in this book extend to the potential societal impacts of AI, particularly in areas like privacy, bias, and decision-making transparency. For example, ethical AI frameworks are necessary to ensure that AI applications do not perpetuate existing biases or infringe on individual privacy rights. The book also highlights the importance of integrating human intuition in AI systems to account for the nuances of human judgment that purely data-driven models might overlook.

CHAPTER 14

Harnessing Predictive Foresight and Human Intuition

Integrating AI into the heart of organizational strategy involves a profound shift that transcends mere technological adoption and means redefining the essence of decision-making, innovation, and value creation. This transformative path requires recognizing that the fusion of predictive analytics and human intuition is critical to unlocking unparalleled potential.

In a world increasingly dominated by data-driven decision-making, there is a growing recognition of its inherent limitations. Purely quantitative approaches, while valuable, risk sidelining the nuanced, creative, and often unpredictable elements of human insight that drive true innovation. The teachings in Vishen Lakhiani's *The Buddha and the Badass* and *The Code of The Extraordinary Mind* emphasize the untapped power of human consciousness when harmonized with AI's capabilities. There is a path forward where AI and human intuition merge to create a new paradigm of excellence and insight.

The transition is more than technological; it is a cultural and philosophical renaissance that touches every aspect of business operations. From the early struggles to gain executive buy-in for AI initiatives to witnessing the transformative impact of early AI models on supply chain efficiency, each step is a lesson in resilience, innovation, and the power of visionary leadership.

The future of work is intricately connected with the evolving landscape of data and AI. This book explores how data and AI are tools and integral components of a transformative shift in approaching work, decision-making, and innovation. By blending predictive analytics with human intuition, organizations can create a more dynamic and resilient workforce better equipped to navigate the complexities of the modern business environment.

The hurdles encountered—talent gaps, ethical problems, and resistance to change—are but waypoints on a journey of discovery and growth. They remind us of the importance of continuous learning, the need to confront and overcome biases, and the remarkable adaptability of humans and organizations when faced with the unknown.

The heart of transitioning to an AI-first approach lies in harmonizing AI's predictive capabilities with the irreplaceable value of human intuition. What is required is a strategic framework for embedding AI into organizational culture—not as a replacement for human insight but as its extension and enhancement. Pilot programs, strategic scaling, and an unwavering commitment to ethical considerations are the foundational elements of such a transition.

The Synergy of Predictive Analytics and Human Intuition

AI and ML algorithms are increasingly making decisions that affect every aspect of our lives. The question of where human intuition fits into this new paradigm has never been more relevant. When navigating the vast sea of data, it is crucial to recognize that the richness of human insight cannot be fully replicated or replaced by algorithms.

The Limitations of Data-Driven Decisions

While data-driven decision-making has transformed industries by providing insights at an unprecedented scale, it has limitations. A purely quantitative approach often fails to capture the full spectrum of human experience, missing nuances that drive behavior, creativity, and innovation. For instance, algorithms can predict consumer behavior based on past trends but overlook emerging cultural shifts that have not yet been captured in the data. Such blind spots can lead to decisions that, while optimized against historical patterns, are misaligned with the current or future state of human affairs.

Real-world examples abound in which data-driven models have faltered, failing to predict outliers or unprecedented events. The financial models leading up to the 2008 financial crisis are a case in point; they could not foresee a market collapse because they were based on historical data that excluded such a possibility. This underscores the critical need for human intuition—a gut feeling or an experienced-based hunch—that can anticipate changes not yet seen in the data or interpret the implications of data in ways that algorithms cannot.

Enhancing Human Intuition with AI

The true potential of AI lies not in replacing human intuition but in augmenting it. AI can process and analyze vast datasets far beyond human capability, which, when integrated with human intuition, can fuel innovative problem-solving, creative thinking, and strategic planning.

Consider the domain of healthcare. While tools can highlight areas of concern with remarkable accuracy, the final judgment often lies with the healthcare professional who integrates this information—using their medical expertise, understanding of the patient's history, and even the patient's nonverbal cues. The synergy between AI and human intuition can lead to more accurate diagnoses, personalized treatment plans, and better patient outcomes.

Moreover, AI can catalyze creativity. In design and content creation, AI tools can generate countless ideas and variations, which creative professionals can refine and iterate upon. Such a collaborative process between AI and human creators opens up new possibilities for innovation that push the boundaries of what can be imagined and achieved.

Looking to the future, integrating AI into decision-making processes offers a path to enhancing abilities and intuition. By recognizing the limitations of a solely data-driven approach and embracing the potential for AI to extend capabilities, we can forge a future where technology and humanity advance hand in hand.

The path forward is not about choosing one over the other but finding the right blend.

Preparing for the Unseen: Fostering Resilience in an AI-Driven World

In navigating the transition to AI-first organizations, an often overlooked yet crucial component is the capacity to prepare for the unseen—those unpredictable, transformative events that lie beyond the horizon of current data and algorithms.

Navigating Unpredictability with Predictive Analytics

Powered by AI and ML, predictive analytics can sift through historical data and identify patterns that forecast future outcomes. These insights can be invaluable in strategic planning, risk management, and resource allocation. However, the true challenge arises when past data may not accurately represent future possibilities—such as unprecedented global events or rapid technological advancements. Human intuition becomes paramount to navigating uncharted waters where data may not yet exist.

Consider the rapid rise of blockchain technology and its disruptive impact on various industries. Early on, predictive models based on traditional financial systems could not foresee the paradigm shift blockchain would introduce. The visionary foresight of some leaders and innovators, coupled with emerging data trends, recognized and capitalized on the opportunities blockchain presented, demonstrating the power of blending predictive analytics with human insight.

The Role of Human Intuition in Steering AI

Human intuition, the innate ability to understand or know something without conscious reasoning, plays a critical role in interpreting and acting upon the insights derived from predictive analytics. It allows leaders to consider factors beyond the data—such as ethical implications, societal trends, and environmental considerations—to ensure that decisions are data-informed and holistically aligned with broader organizational values and societal good.

For example, AI-driven models might identify significant cost-reduction opportunities by automating certain job functions. However, intuitive leadership recognizes the broader implications of such a move, including employee morale, company culture, and brand reputation. By integrating such considerations into decision-making, organizations can navigate the complexities of an AI-driven world with empathy and responsibility to ensure that technological advancement enhances rather than detracts from human values.

Cultivating Organizational Resilience

The unpredictable nature of our future, particularly in an era dominated by rapid technological change, calls for an organizational culture that values resilience and adaptability. This will foster an environment where continuous learning, open-mindedness, and the capacity to pivot are ingrained in an organization's DNA. Encouraging teams to experiment, embrace failures as learning opportunities, and continuously question and refine assumptions are vital to building resilience.

Moreover, promoting interdisciplinary collaboration within organizations can uncover unique insights and innovative solutions that a siloed approach might miss. Drawing on diverse perspectives and expertise enhances the organization's ability to anticipate challenges and adapt to change, ensuring long-term sustainability and success.

Fostering an AI-First Culture: A Leadership Imperative

As organizations embark on the journey to become AI-first, a critical factor is leadership's ability to foster a culture that embraces AI as a transformative force. This cultural shift is not merely about adopting new technologies; it requires a fundamental change in mindset, values, and behaviors across the organization.

Cultivating an AI-first culture is imperative for leadership in the age of AI. This involves creating an environment where experimentation, continuous learning, and collaboration between humans and machines are not just encouraged but the norm. Leaders must champion the realization of AI as a tool for augmenting human capabilities, driving innovation, and creating value. They must inspire their teams to embrace the opportunities AI presents while also addressing the fears and uncertainties that inevitably accompany such a profound transformation.

Organizational Structures for an AI-First Future

As organizations transition to an AI-first approach, it is critical to consider how their structures and processes must evolve to support this new paradigm. With siloed departments and rigid decision-making chains, traditional hierarchical models may need help to keep pace with the speed and fluidity required in an AI-driven world.

Organizations should move towards more agile, networked structures that enable cross-functional collaboration and rapid experimentation. This may involve creating dedicated AI teams that work across the organization, embedding AI expertise within each business unit, or establishing internal "AI centers of excellence" that provide guidance and support.

Such structures should facilitate close interaction between human decision-makers and AI systems. Regular touchpoints, feedback loops, and opportunities for human oversight can help ensure that AI insights are being interpreted and applied in context-appropriate ways.

Moreover, as AI automates specific tasks and decision-making processes, organizations must rethink the roles and responsibilities of their human workforce. This may require redesigning jobs to focus on higher-level strategic tasks and providing upskilling opportunities to help employees adapt to working alongside AI.

One critical step in becoming an AI-first organization is balancing leveraging AI for efficiency and scalability while leaving room for human intuition and judgment. Overreliance on AI can lead to decisions optimized for specific metrics that do not account for broader contexts or unintended consequences.

To achieve balance, organizations should develop clear guidelines and protocols for when and how AI systems are used in decision-making. This might involve setting thresholds that require human review or establishing human-in-the-loop processes where AI provides recommendations, but people make final decisions.

Challenges and Solutions in the Journey to AI-First

The path to becoming an AI-first organization has its challenges. Leaders must navigate a complex landscape of technical, cultural, and ethical considerations to integrate AI into operations and decision-making processes successfully. **Some key challenges include:**

- **Skill gaps**: As AI technologies advance rapidly, many organizations need help finding and retaining talent with the necessary technical skills. This can hinder the development and implementation of AI systems. Organizations must invest heavily in upskilling and reskilling their existing workforce to address this. Partnering with educational institutions, providing in-house training programs, and fostering a culture of continuous learning can help bridge the skill gaps.

- **Resistance to change**: Adopting an AI-first approach often requires significant organizational structure, processes, and role changes. Employees may feel threatened by the prospect of AI automating parts of their jobs or may be hesitant to trust AI-driven decisions. Overcoming such resistance requires clear communication from leadership about the benefits of AI and how it will augment rather than replace human capabilities. Engaging employees in the AI journey, providing opportunities to collaborate with AI systems, and celebrating successes can help build buy-in and enthusiasm.

- **Data quality and governance**: AI systems are only as good as the data they are trained on. Organizations often face challenges with data silos, inconsistencies, and lack of quality control. Investing in robust data governance frameworks, data integration tools, and data literacy training for employees is crucial. Establishing clear data ownership, maintaining data hygiene, and ensuring compliance with data privacy regulations are all critical components of a successful AI strategy.

- **Ethical concerns**: As AI systems become more autonomous and influential in decision-making, organizations must grapple with complex moral questions. How can we ensure that AI systems are fair, unbiased, and transparent? How do we hold AI accountable for its decisions? What are the societal implications of AI adoption? Organizations must develop solid ethical frameworks for AI development and use them to navigate such concerns. This includes establishing oversight committees, conducting regular audits and impact assessments, and engaging in open dialogue with stakeholders about the ethical implications of AI.

- **Integration complexity**: Integrating AI systems into IT infrastructures and business processes can be a significant technical challenge. Legacy systems, data compatibility issues, and the need for real-time processing can create hurdles. Organizations must take a strategic approach to AI integration, starting with pilot projects and gradually scaling up. Investing in flexible, modular architectures and partnering with experienced AI vendors can help streamline the integration process.

Despite such challenges, the potential benefits of becoming an AI-first organization are immense. AI can help organizations achieve unprecedented efficiency, insight, and competitive advantage by augmenting human capabilities, driving innovation, and enabling data-driven decision-making.

The key to success lies in approaching the AI journey with a spirit of experimentation, collaboration, and continuous learning. Leaders must be willing to take calculated risks, learn from failures, and adapt strategies as the technology and business landscape evolves.

By fostering an AI-first culture, investing in the right skills and tools, and prioritizing ethical considerations, organizations can not only overcome the challenges of AI adoption but also position themselves at the forefront of the AI revolution. The journey may be complex, but the destination—a future where human and machine intelligence work together seamlessly to solve our most significant challenges—is well worth the effort.

Additionally, organizations should invest in tools and interfaces that make AI systems more transparent and explainable to human users. This can help build trust in AI insights while empowering humans to spot potential biases or errors.

Strategies and Tools for Harmonizing AI and Human Intelligence

Integrating AI with human intuition effectively requires a thoughtful approach to change management and the correct set of enabling technologies. **Some key strategies and tools include:**

- **Collaborative platforms**: Invest in AI-powered platforms that facilitate collaboration and knowledge-sharing across the organization. These can help surface relevant insights and expertise while enabling humans to provide feedback and refine AI models.

- **Explainable AI**: Prioritize AI systems that clearly explain their outputs and decisions. This transparency is critical for building trust and allowing humans to interpret AI insights based on their knowledge and experience.

- **Continuous learning**: Establish programs and incentives to encourage ongoing learning and skill development in technical AI skills and the "soft skills" required for effective human-machine collaboration.

- **Governance frameworks**: Develop clear administrative structures and ethical frameworks to guide the use of AI, ensuring that it aligns with organizational values and priorities. Regular audits and impact assessments can help surface and address any unintended consequences.

- **Experimentation and iteration**: Foster a culture of rapid improvement that recognizes that becoming an AI-first organization requires continuous learning and adaptation.

By embracing these strategies and tools, organizations can work towards a future where AI and human intelligence are seamlessly integrated to drive efficiency, innovation, creativity, and expanded human potential.

The Conscious Use of AI: Navigating Ethical and Societal Implications

Understanding the responsibility of wielding such powerful technology is crucial to ensuring that advancements in AI drive economic and operational efficiency and contribute to society.

As organizations march toward an AI-driven future, the conversation inevitably shifts toward this transformative technology's ethical and societal implications. The conscious use of AI demands a thorough examination of how these tools impact the bottom line, the broader tapestry of society, and the ethical frameworks that govern our decisions.

One of the most significant challenges I faced in AI implementation involved a project for credit scoring at a central bank. Credit scoring is a critical function in banking, determining who gets access to credit and on what terms. However, it is also an area fraught with ethical concerns, particularly bias and fairness.

During the project's initial stages, we discovered that the historical data used for credit scoring, even before the introduction of AI, had inherent biases. These biases were largely invisible in traditional models but had real-world impacts on decision-making. For instance, certain demographic groups consistently received lower credit scores not because of their creditworthiness but because of systemic issues in how data had been collected and used over the years.

The challenge was clear: if AI and ML techniques were applied to this biased data, we would perpetuate and possibly even exacerbate these biases. The risk was not just technical but deeply ethical. We needed to ensure that the AI systems we developed would be fair, transparent, and accountable. **To tackle these challenges, we focused on several key areas:**

- **Bias detection and mitigation**: We implemented algorithms designed to detect bias in the data and the resulting models. One approach involved using fairness metrics to evaluate our AI model outputs, ensuring no demographic group was disproportionately disadvantaged.

- **Data parity and noise addition**: To mitigate the effects of bias, we introduced techniques such as data parity and adding random noise to the data. Data parity involved ensuring that our training data was balanced across different demographic groups, while noise addition helped prevent the model from learning biases present in the data.

- **Human oversight**: Recognizing that no algorithm is perfect, we established a process of continuous human oversight. This included regular audits of the AI system's outputs, where data scientists and ethicists reviewed decisions made by the AI to identify any potential ethical concerns.

- **Transparency and explainability**: We also prioritized transparency and explainability in our AI models. This meant developing models that were not only accurate but also interpretable, allowing stakeholders to understand how decisions were being made. This was crucial for building trust within the bank and among its customers.

- **Outcomes and lessons learned:** Implementing these ethical safeguards significantly improved the fairness of the credit scoring system. The bank saw reduced bias-related discrepancies in credit scores, leading to more equitable access to credit across different demographic groups. Furthermore, the AI models' transparency and explainability helped build trust among stakeholders, ensuring that the system was practical and ethically sound.

This experience reinforced the critical importance of integrating ethical considerations into AI development. It taught me that AI is not just about technical performance but about ensuring that technology serves the broader goals of fairness and justice. This project also highlighted the need for continuous monitoring and adaptation, as ethical challenges in AI are not static but evolve.

Ethical AI: A Foundation for Trust and Integrity

At the heart of ethical AI lies the commitment to developing and deploying technologies that are fair, transparent, accountable, and devoid of bias. This commitment is the foundation of building trust between organizations, customers, and society. However, achieving this ideal is fraught with challenges, from the inherent biases in training data to the opaque nature of some AI algorithms.

To navigate these challenges, organizations must adopt robust ethical frameworks that guide the development and deployment of AI. These should include mechanisms for regular ethical audits, transparency reports, and channels for addressing grievances. For example, IBM's AI Ethics Board oversees its ethical AI practices to ensure that all AI applications align with its Principles of Trust and Transparency. Such oversight bodies can be crucial in maintaining ethical standards and fostering public trust.

The Societal Impact of AI: Beyond the Bottom Line

As AI technologies become more integrated into various facets of life, their societal impact becomes an increasingly critical consideration. From reshaping the labor market to influencing public opinion, AI's ripple effects are profound. Therefore, organizations must consider these broader implications and ensure that AI initiatives contribute positively to society.

One area of particular concern is the potential for AI to exacerbate inequality, whether through job displacement or by reinforcing existing biases. Proactive measures, such as investing in workforce retraining programs and ensuring diversity in AI development teams, can help mitigate these risks. Additionally, engaging with stakeholders from various sectors to discuss and address the societal implications of AI can lead to more inclusive and beneficial outcomes.

Navigating the Ethical Landscape Collaboratively

The complexity of ethical AI usage necessitates a shared approach involving dialogue and partnership between industry, academia, government, and civil society. By pooling knowledge, resources, and perspectives, stakeholders can develop comprehensive guidelines and standards for ethical AI that reflect a wide range of interests and concerns.

Furthermore, organizations can leverage such collaborations to stay abreast of emerging ethical challenges and innovative solutions. For instance, participation in initiatives like the Partnership on AI or the IEEE Global Initiative on Ethics of Autonomous and Intelligent Systems enables companies to contribute to and benefit from collective efforts to address ethical considerations in AI.

Navigating AI's ethical and societal implications is an ongoing journey demanding constant vigilance, reflection, and adjustment. As we delve into these complexities, it becomes clear that integrating AI into our organizations and societies is not just a technological endeavor but a profoundly human one.

Harmonizing AI with Human Intelligence: The Future of Organizational Decision-Making

The fusion of AI with human intelligence represents a frontier in the evolution of organizational decision-making. This convergence is not about replacing human intuition with artificial algorithms but creating a symbiotic relationship where each complements the other. As we venture into this future, it becomes imperative for organizations to understand how to harmonize these distinct yet interconnected forms of intelligence.

Augmenting Human Intelligence with AI

AI brings an unparalleled capacity for processing vast amounts of data, identifying patterns, and predicting outcomes with a degree of previously unattainable accuracy. However, AI needs more nuanced understanding, emotional intelligence, and creative problem-solving—something humans excel at. The challenge and opportunity for organizations lie in leveraging AI to augment human intelligence to enhance decision-making with a blend of analytical prowess and human insight.

For instance, companies like Google have utilized AI to augment human creativity in developing products and solutions. Through tools like TensorFlow (Google's open-source ML platform), developers and researchers can experiment with AI algorithms to enhance their work, from refining search algorithms to improving energy consumption efficiency in data centers.

Ethical Considerations in the AI-Human Partnership

Ethical considerations take on a new dimension when navigating this integration. The partnership between AI and human intelligence must be governed by principles that ensure fairness, accountability, and transparency. This includes addressing the potential biases AI systems may inherit from their human creators or the data they are trained on.

Initiatives such as the AI Now Institute, which focuses on the social implications of artificial intelligence, can provide organizations with guidance on best practices. Companies can also engage in such conversations to develop AI that respects and enhances our shared human values.

Preparing for a Future Shaped by AI and Human Synergy

The path to harmonizing AI with human intelligence requires a concerted effort to develop the skills and frameworks needed for this new era of decision-making. This involves technical training in AI and ML and fostering the soft skills that AI cannot replicate, such as emotional intelligence, ethical reasoning, and creative thinking.

Programs like IBM's P-TECH—a public-education model that provides young people with the technical and professional skills required in the digital era—represent steps in the right direction. Similarly, organizations should invest in continuous learning and development programs that empower employees to thrive in a landscape where AI is a tool for amplification, not replacement.

The journey towards integrating AI and human intelligence in organizational decision-making is both exciting and challenging. It promises a future where the analytical power of AI, combined with the creative and ethical capacities of human intelligence, drives innovation and ethical decision-making to new heights. The overarching theme of extending our cognitive capabilities is the fusion of technology with profound human values and instincts, which will pave the way for a future that is not only technologically advanced but also profoundly human.

Case Studies: Bridging AI and Human Insight

Unilever's Marketing Strategies

Unilever leverages AI to enhance its marketing strategies, combining data-driven insights with creative human input (Texta n.d.; AI Content Factory n.d.; Accenture 2023; Unilever 2024a, 2024b).

- **AI capabilities**: AI tools analyze consumer data, market trends, and social media sentiment to identify target audiences and predict campaign effectiveness.

- **Human intuition**: Marketing teams at Unilever use their creativity and understanding of brand messaging to develop and refine marketing campaigns based on AI insights.

- **Outcome**: This collaboration has led to more effective and engaging marketing campaigns, higher ROI, and stronger brand loyalty. For instance, Unilever's Dove Real Beauty campaign effectively utilized AI insights to connect with consumers emotionally, resulting in widespread acclaim and increased sales.

Stitch Fix: Merging Data Science with Human Stylists

Stitch Fix, an online personal styling service, leverages AI to analyze customer preferences and trends, providing data-driven recommendations for clothing (WBR Insights n.d.; Maskalevich n.d.; Allsup 2023; Ransbotham 2024; Stitch Fix and n.d.; La-Gaffe 2023).

- **AI capabilities**: The company's algorithms analyze data from customer surveys, feedback, and purchase history to generate personalized clothing recommendations.

- **Human intuition**: Human stylists review the AI-generated suggestions, adding a personal touch based on their experience and understanding clients' needs and preferences.

- **Outcome**: This combination ensures high customer satisfaction and retention by blending AI's efficiency with human stylists' empathy and personal insights. Stitch Fix has seen significant success, with its personalized approach leading to high customer engagement and loyalty.

These examples illustrate the powerful synergy between AI and human intuition across various industries, demonstrating how organizations can harness their strengths to achieve remarkable outcomes.

Charting a Future Where AI Complements Human Wisdom

The future of innovation and decision-making lies not in choosing between AI and human wisdom—but in harmonizing them. AI-first organizations are evolving from a narrative of replacement to one of augmentation and partnership.

AI is not merely a tool but an extension of our being. Thought leaders like Joe Dispenza and Vishen Lakhiani all point towards a profound truth: AI, when aligned with our values and purpose, can amplify our intuition and unlock new realms of human potential.

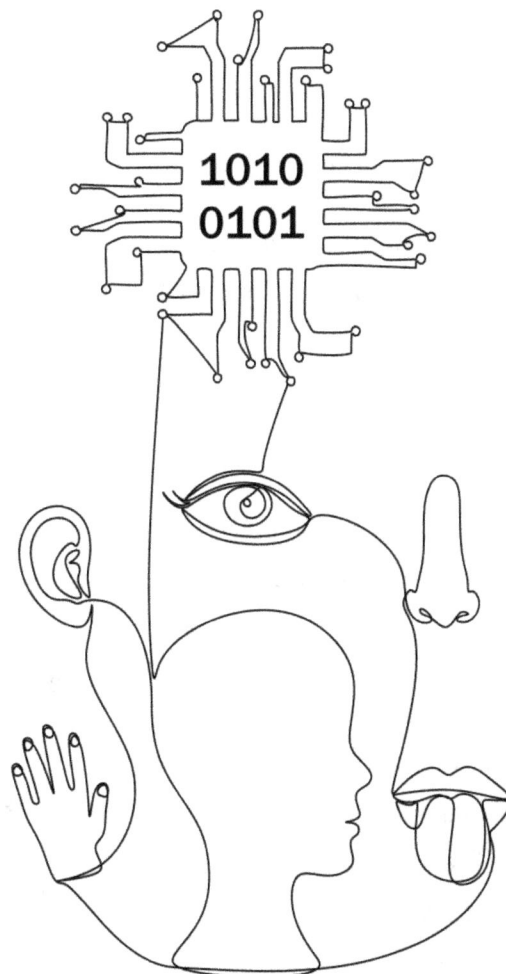

Figure 47. Extending Perception, Enhancing Insight in the Digital Age

AI becomes a receiver and amplifier of human consciousness, a conduit to explore uncharted territories of creativity, innovation, and self-discovery. By melding our intuitive wisdom with the computational power of AI, we give rise to a new form of intelligence not constrained by the limitations of either humans or machines.

AI as a Receiver of Consciousness

When aligned with our core values and ethical frameworks, AI is a tool that can magnify our ability to solve complex problems, create limitlessly, and make decisions informed by a deep understanding of data and the human experience. As Joe Dispenza eloquently teaches, by tapping into our consciousness and melding it with AI's capabilities, we unlock possibilities where intuition and intelligence can coalesce to guide us toward more enlightened choices and innovations.

Charting the path forward, **the call to action for leaders, innovators, and thinkers is clear:**

- **Cultivate a mindset of augmentation**: View AI as a partner in the creative process, a tool that augments human capabilities rather than replaces them.

- **Invest in ethical frameworks**: Develop and adhere to ethical frameworks that ensure AI is used to enhance societal well-being, respect privacy, and promote equity.

- **Foster a culture of continuous learning**: Encourage an organizational culture that values learning, adaptability, and the exploration of new frontiers at the intersection of technology and human potential.

- **Embrace interdisciplinary collaboration**: Promote partnerships across disciplines that bring together technologists, artists, ethicists, and business leaders to create AI applications that reflect the richness of human experience.

Conclusion

Reflecting on the transformative journey towards AI-first organizations, it becomes clear that the path is not just about technological adoption but a fundamental shift in how we approach decision-making, innovation, and value creation. We can unlock unprecedented insight, efficiency, and competitive advantage by harnessing the symbiotic relationship between human intuition and artificial intelligence.

However, the road to AI-first is not without its obstacles. Organizations must navigate a complex landscape of challenges, from bridging skill gaps and overcoming resistance to change to ensuring data quality and grappling with ethical concerns. The key to success lies in approaching these challenges with a spirit of experimentation, collaboration, and continuous learning.

By investing heavily in upskilling and reskilling initiatives, fostering a culture of lifelong learning, and partnering with educational institutions, organizations can build the talent pipeline necessary for AI success.

Engaging employees in the AI journey, providing opportunities for human-machine collaboration, and celebrating successes can help overcome resistance and build enthusiasm for this new working method.

Effective AI systems require robust data governance frameworks, clear data ownership, and a commitment to data hygiene and privacy. Organizations must also develop solid ethical frameworks, establish oversight, conduct regular audits, and engage in open dialogue about AI's societal implications.

To navigate the technical complexities of AI integration, a strategic approach to balancing pilot projects with scalable architectures is critical. Partnering with experienced AI vendors and prioritizing flexible, modular systems can help streamline the journey.

Ultimately, the AI-first path is one of continuous learning and adaptation. By cultivating a culture of experimentation, investing in the right skills and tools, and always keeping human values at the center, we can overcome the challenges of AI adoption and redefine what is possible for organizations and society.

Standing on the precipice of this new era, let us move forward with courage, curiosity, and a deep commitment to harnessing AI's power to benefit all. The journey ahead may be complex, but the destination—a future where human and machine intelligence work together seamlessly to solve our most significant challenges—is well worth the effort.

Ultimately, becoming an AI-first organization is not just a business imperative but an opportunity to shape a future that is not only technologically advanced but also profoundly human-centric. By staying true to our values, ethics, and the irreplaceable spark of human ingenuity, we can ensure that the story of AI is ultimately a story of empowerment, purpose, and limitless potential for good.

Acknowledgments

This book has been a journey of collaboration, and I am deeply grateful to everyone who contributed to its creation.

First and foremost, I would like to thank Peter Letzelter-Smith, whose expertise in developmental, line, and copy editing brought clarity and precision to the manuscript. Your meticulous attention to detail and insightful feedback helped shape this book into what it is today.

I also extend my heartfelt thanks to Dr. Magda Wojcik for her invaluable contributions to developmental editing. Your guidance and thoughtful suggestions were instrumental in refining the book's structure and flow.

A special thanks go to Giuliana Magnotti and Fitria Alfa Chasana for their stunning line art, which beautifully complements the themes of this book. Your artistic vision helped bring complex ideas to life in a visual form.

I am immensely grateful to Suman Al Mamun and his team for their exceptional work formatting the book. Your skillful design and formatting ensured that every element was cohesive and professional, ready for publication.

Lastly, I would like to acknowledge the various AI tools that assisted me throughout this process, from image generation to text editing and research. These tools provided crucial support in enhancing the book's creative and technical aspects.

Thank you for your dedication and contributions. This book would not have been possible without your help.

Reference List

3Pillar Global. (2021, September 7). *Customer Experience Case Studies*. https://www.3pillarglobal.com/insights

7T. (n.d.). *Digital transformation failure examples: Causes of failed AI projects and lessons learned*. Retrieved May 10, 2024, from https://7t.co/blog/digital-transformation-failure-examples-causes-of-failed-ai-projects-and-lessons-learned/

Accenture. (2021, March 30). *Accenture launches 360-degree value reporting experience*. https://newsroom.accenture.com/news/accenture-launches-360-degree-value-reporting-experience.htm

Accenture. (2023, December 5). Unilever and Accenture collaborate on next generation AI. https://newsroom.accenture.com/news/2023/unilever-and-accenture-collaborate-on-next-generation-ai

Ackoff, R. (1989). From data to wisdom. *Journal of Applied Systems Analysis*, 16, 3–9. https://www.scirp.org/reference/ReferencesPapers?ReferenceID=713373

Adobe. (2024, March 26). *Adobe accelerates data-driven personalization at scale with Adobe Experience Platform innovations*. https://news.adobe.com/news/news-details/2024/Adobe-Accelerates-Data-Driven-Personalization-at-Scale-with-Adobe-Experience-Platform-Innovations/

Advisory Board. (n.d.). *CEO Warner Thomas' vision for digital transformation at Sutter Health.* Retrieved May 10, 2024, from https://www.advisory.com/topics/strategy-planning-and-growth/2023/07/digital-transformation-at-sutter-health

AIContentfy. (2024, May 10). *10 effective strategies for reducing costs and maximizing ROI*. https://aicontentfy.com/en/blog/strategies-for-reducing-costs-and-maximizing-roi

AI Content Factory. (n.d.). *AI content case studies*. Retrieved July 30, 2024, from https://theaicontentfactory.com/ai-content-case-studies/

Al Farabi, H. (2023a, August 26). *From air mattresses to a global phenomenon: Airbnb case study*. LinkedIn. https://www.linkedin.com/pulse/from-air-mattresses-global-phenomenon-airbnb-case-study-farabi

Al Farabi, H. (2023b, September 9). *Case study: Spotify; changing the tune of the music industry*. LinkedIn. https://www.linkedin.com/pulse/case-study-spotify-changing-tune-music-industry-hossain-al-farabi

Alberdi, R. (n.d.). *(Masterclass) The Hidden Secrets Behind Netflix's Success*. ThePower Business School. Retrieved May 10, 2024, from https://www.thepowermba.com/en/blog/netflix-success

Alexander S. (2023, April 19). *Tesla's use of AI: A revolutionary approach to car technology*. LinkedIn.

https://www.linkedin.com/pulse/teslas-use-ai-revolutionary-approach-car-technology-alexander-stahl

Allsup, Maeve. (2023, April 3). How Stitch Fix uses AI to take personalization to the next level." Retail Brew. https://www.retailbrew.com/stories/2023/04/03/how-stitch-fix-uses-ai-to-take-personalization-to-the-next-level

Almalki, I. (2020, October 15). *Yahoo: The story of strategic mistakes*. The Strategy Story. https://thestrategystory.com/2020/10/15/yahoo-the-story-of-strategic-mistakes/

Alsumidaie, M. (2024, May 23). *AI's astonishing impact on drug development and ethics*. Clinical Trial Vanguard. Accessed July 10, 2024. https://www.clinicaltrialvanguard.com/conference-coverage/ais-astonishing-impact-on-drug-development-and-ethics

Althris Training. (2023, December 15). *The Spotify model: A symphony of agile scaling, autonomy, and continuous improvement*. LinkedIn. https://www.linkedin.com/pulse/spotify-model-symphony-agile-scaling-autonomy-continuous-improvement-anzge

Alves, J. (2020, April 3). *How experience-led growth catapulted Airbnb to unicorn status*. Chattermill. https://chattermill.com/blog/airbnb-customer-experience

Amazon Web Services. (2023). *Gilead's journey from migration to innovation on AWS*. https://aws.amazon.com/solutions/case-studies/gilead-data-case-study/

Anderson, R. (2021, July 29). *Average data breach costs reach record level of $4.24 million per breach*. NetSec News. https://www.netsec.news/average-data-breach-costs-reach-record-level-of-4-24-million/

Andrews, L., & Bucher, H. (2022). Automating discrimination: AI hiring practices and gender inequality. *Cardozo Law Review 44*(1). https://cardozolawreview.com/automating-discrimination-ai-hiring-practices-and-gender-inequality/

Architech. (2024, February 1). *Building data architecture for AI/ML use cases*. Medium. https://medium.com/@architechcro/building-data-architecture-for-ai-ml-use-cases-ade541f110a8

Asif, H. (2023, September 29). *Monolithic vs microservices architecture: Case study of Netflix and Atlassian*. LinkedIn. https://www.linkedin.com/pulse/monolithic-vs-microservices-architecture-case-study-netflix-asif

AstraZeneca. (n.d.). *AstraZeneca Data and AI Ethics*. Accessed July 10, 2024. https://www.astrazeneca.com/sustainability/ethics-and-transparency/data-and-ai-ethics.html

Atlan. (2023, August 1). *Netflix Metacat: Origin, architecture, features & more*. https://atlan.com/metacat-netflix-open-source-metadata-platform/

AWS. (2016). *Netflix case study*. https://aws.amazon.com/solutions/case-studies/netflix-case-study/

AWS. (n.d.-a). *Amazon adopts Amazon Aurora for inventory database*. Retrieved May 10, 2024, from https://aws.amazon.com/solutions/case-studies/amazon-fulfillment-aurora/

AWS. (n.d.-b). Capital One on AWS. Retrieved May 10, 2024, from https://aws.amazon.com/solutions/case-studies/capital-one/

B4AD56. (2018, January 31). *A blockbuster failure and the changing media landscape*. Harvard Business School: Digital Initiative. https://d3.harvard.edu/platform-digit/submission/a-blockbuster-failure-and-the-changing-media-landscape/

Barkho, G. (2023, October 25). How the fake meat industry is trying to reinvent itself. *Modern Retail*.

https://www.modernretail.co/marketing/how-the-fake-meat-industry-is-trying-to-reinvent-itself/

Barr, S. (2019, August 27). BBC seeks to increase younger audience through data analytics. *Computer Weekly*. https//www.computerweekly.com/news/252456977/BBC-seeks-to-increase-younger-audience-through-data-analytics

BBC Datalab. (n.d.). *Machine learning at the BBC*. Accessed July 10, 2024. https://datalab.rocks/

Bean, R. (2021). *Fail fast, learn faster: Lessons in data-driven leadership in an age of disruption, big data, and AI*. Wiley.

Bendor-Samuel, P. (2020, October 9). IBM splits into two companies. *Forbes*. https://www.forbes.com/sites/peterbendorsamuel/2020/10/09/ibm-splits-into-two-companies/?sh

Benton, J. (2010, August 5). *How the Guardian Is pioneering data journalism with free tools*. NiemanLab. https://www.niemanlab.org/2010/08/how-the-guardian-is-pioneering-data-journalism-with-free-tools/

Berg, K. (2020). *Spotify's people strategy*. Spotify HR Blog. https://hrblog.spotify.com/2020/11/25/spotifys-people-strategy

Bhambhri, A. (2024, March 26). *New AI integrations in Adobe Experience Platform*. Adobe Blog. https://blog.adobe.com/en/publish/2024/03/26/new-ai-integrations-in-adobe-experience-platform

BioSpace. (2021, September 13). *Biopharma AI collaborations driving innovative change in drug development*. Accessed July 10, 2024. https://www.biospace.com/article/biopharma-ai-collaborations-driving-innovative-change-in-drug-development-

Bivins, S. S. (2014). A transformational change at IBM. *PMI® Global Congress 2014, Phoenix, AZ*. Project Management Institute. https://www.pmi.org/learning/library/transformational-change-ibm-9297

Bjelland, O. M., & Wood, R. C. (2008, October 1). An inside view of IBM's "innovation jam." *MIT Sloan Review*. https://sloanreview.mit.edu/article/an-inside-view-of-ibms-innovation-jam/

Bluefin. (2023, July 27). *Data breaches: Record high costs and mitigation solutions*. https://www.bluefin.com/bluefin-news/data-breaches-record-high-costs-mitigation-solutions/

Bonnington, C. (2012, September). Tim Cook apologizes for Maps. *Wired*. https://www.wired.com/2012/09/tim-cook-apologizes-for-maps/

Bounegru, L. (n.d.). *Data journalism at the BBC*. DataJournalism.com. Accessed July 10, 2024. https://datajournalism.com/read/handbook/one/in-the-newsroom/data-journalism-at-the-bbc

Boyd, S. (2019, July 30). *Spotify: An emergent organization*. Medium. https://medium.com/work-futures/spotify-an-emergent-organization-df9da1125c

Brand Credential. (2023, December 8). *Netflix marketing strategy: Streaming success*. https://www.brandcredential.com/post/netflix-marketing-strategy-streaming-success

BrightTALK. (2021, March 17). *Case study in DataOps: How data puts the game in gaming*. https://www.brighttalk.com/webcast/15913/475399

Brown, N. (2023, April 26). *How AstraZeneca is applying AI, imaging, data analytics (AI-Driven Drug Development Summit Europe 2023)*. SlideShare. Accessed July 10, 2024. https://www.slideshare.net/slideshow/how-astrazeneca-is-applying-ai-imaging-data-analytics-aidriven-drug-development-summit-europe-2023/257577945

Bulah, B. M., et al. (2023). Institutional work as a key ingredient of food innovation success: The case of plant-based proteins. *Environmental Innovation and Societal Transitions* 47: 100697.

https://doi.org/10.1016/j.eist.2023.100697.

Burg, D., & Ausubel, J. H. (2021, August 18). Moore's Law revisited through Intel chip density. *PLOS One*, 15(7), e0235993. https://journals.plos.org/plosone/article?id=10.1371/journal.pone.0256245

Burgess, M. (2018, September 7). The British Airways hack is impressively bad: BA is the latest company to be hit by hackers; We chart the biggest data breaches of 2018. *Wired*. https://www.wired.com/story/hacks-data-breaches-in-2018/

Caldwell, C., & Liu, J. (2021, March 10). *Dominance of WeChat Pay and Alipay in the Chinese digital payments industry*. Focus Finance. https://www.focusfinance.org/post/dominance-of-wechat-pay-and-alipay-in-the-chinese-digital-payments-industry

California v. Sutter Health System. 130 F. Supp. 2d 1109 (N.D. Cal. 2001). https://casetext.com/case/california-v-sutter-health-system

Campbell, K. (2024, March 28). Taking a data-first approach to digital transformation. *Forbes*. https://www.forbes.com/sites/forbestechcouncil/2024/03/28/taking-a-data-first-approach-to-digital-transformation/

Castus. (2021, October 5). *The downfall of Sears: 5 key reasons why the retail giant went under*. https://www.castusglobal.com/insights/the-downfall-of-sears-5-key-reasons-why-the-retail-giant-went-under

Catmull, E. (2008, September). How Pixar fosters collective creativity. *Harvard Business Review*. https://hbr.org/2008/09/how-pixar-fosters-collective-creativity

CBS News. (2019, November 11). *Apple Card accused of gender discrimination in its algorithm*. https://www.cbsnews.com/video/apple-card-accused-of-gender-discrimination-in-its-algorithm/

Cecco, L. (2024, February 16). Air Canada ordered to pay customer who was misled by airline's chatbot. *Guardian*. https://www.theguardian.com/world/2024/feb/16/air-canada-chatbot-lawsuit

Chappell, B. (2011, October 10). Netflix kills Qwikster, price hike lives on. *NPR*. https://www.npr.org/sections/thetwo-way/2011/10/10/141209082/netflix-kills-qwikster-price-hike-lives-on

Chatterjee, S. (2024, April 30). *Here's what Google is doing right with their AI strategy, and you should too!* Emeritus. https://emeritus.org/blog/ai-strategy-google/

Chaurasia, N. (2023, September 13). *Netflix marketing strategy*. Sprintzeal. https://www.sprintzeal.com/blog/netflix-marketing-strategy

Chegg. (n.d.). *Case study: UPS deploys routing optimization for big payoff*. Retrieved May 10, 2024, from https://www.chegg.com/homework-help/questions-and-answers/case-study-1201-ups-deploys-routing-optimization-big-payoff-ups-orion-road-integrated-opti-q66413561

Choi, E., Bahadori, M. T., & Schuetz, A. (2015, November 18). Doctor AI: Predicting clinical events via recurrent neural networks. *arXiv:1511.05942*. https://arxiv.org/abs/1511.05942

Chojecki, P. (n.d.). *What is Moore's Law? Is it dead?* Built In. Retrieved May 10, 2024, from https://builtin.com/hardware/moores-law

Chong, D. (2020, April 30). *Deep dive into Netflix's recommender system*. Medium. https://towardsdatascience.com/deep-dive-into-netflixs-recommender-system-341806ae3b48

Christensen, C. M., Bartman, T., & van Bever, D. (2016, September 13). The hard truth about business model innovation. *MIT Sloan Management Review*. https://sloanreview.mit.edu/article/the-hard-truth-about-business-model-innovation/

CIO Influence. (2023, December 14). *NVIDIA's role in utilizing RAG for advanced AI chatbot development in enterprises*. https://cioinfluence.com/it-and-devops/nvidias-role-in-utilizing-rag-for-advanced-ai-chatbot-development-in-enterprises/

Clarke, A. (2024, April 9). *Google fancies a bit of HubSpot: AI's data dilemma*. LinkedIn. https://www.linkedin.com/pulse/google-fancies-bit-hubspot-ais-data-dilemma-adam-clarke-g2ume

Cloud 7 IT Service (2023, October 6). *Case study: Successful cloud security implementation*. LinkedIn. https://www.linkedin.com/pulse/case-study-successful-cloud-security-implementation

Cochran, T. (2021, January 2021). *Maximizing developer effectiveness*. martinFowler.com. https://martinfowler.com/articles/developer-effectiveness.html

Collective Measures. (2021, January 19). *Apple's iOS 14 update*. https://www.collectivemeasures.com/insights/apples-ios-14-update

Convolo. (2024, January 18). *Maximizing ROI: Strategies for increasing return on investment*. https://www.convolo.ai/blog/maximizing-roi-strategies-for-increasing-return-on-investment

Cox, B. (2001, December 27). *Case study: Office Depot gets intelligent with MicroStrategy*. Datamation. https://www.datamation.com/storage/case-study-office-depot-gets-intelligent-with-microstrategy/

Cruth, M. (n.d.). *Discover the Spotify model*. Atlassian. Retrieved May 10, 2024, from https://www.atlassian.com/agile/agile-at-scale/spotify

Culture Partners. (2024, April 16) *Google's company culture: Unveiling organizational values*. https://culturepartners.com/insights/googles-company-culture-unveiling-organizational-values/

Daksh, R. G. (2023, August 31). *Unveiling the streaming revolution: How did Netflix transform from DVD rentals to global streaming*. Medium. https://medium.com/@rgdaksh03122005/unveiling-the-streaming-revolution-how-did-netflix-transform-from-dvd-rentals-to-global-streaming-c1bfc9dcbdd7

Danvers, A. (2023, October 2). *How Zapier uses AI to make automation easy*. Solvaa. https://solvaa.co.uk/how-zapier-uses-ai-to-make-automation-easy-part1/

Data Mesh Learning. (n.d.). *Community*. Retrieved May 10, 2024, from https://datameshlearning.com/community-getting-started/

Datacenters. (2023, October 4). *Maximizing ROI on Dev Ops investments: A comprehensive guide*. https://www.datacenters.com/news/maximizing-roi-on-dev-ops-investments-a-comprehensive-guide.

DataKitchen. (2019, June 29). *The best DataOps articles Q2 2019*. https://datakitchen.io/the-best-dataops-articles-q2-2019/

DataRobot. (n.d.). *Your path to value starts now*. Retrieved May 10, 2024, from https://www.datarobot.com/platform/

Davenport, T. (2017, April 18). What's your data strategy? *Harvard Business Review*. https://hbr.org/webinar/2017/04/whats-your-data-strategy

Davenport, T. (2021, March 12). The future of work now: AI-assisted clothing stylists at Stitch Fix. *Forbes*. https://www.forbes.com/sites/tomdavenport/2021/03/12/the-future-of-work-now-ai-assisted-clothing-stylists-at-stitch-fix/?sh=2772e92b3590

Day, J. (2013, September 27). *FCA fines JPMorgan £137,610,000 for serious trading failings*. Lexology.

https://www.lexology.com/library/detail.aspx?g=44e93ae6-edb6-46c8-8138-97330a4e3d13

De Nitto, F. (2023, June 19). *DataOPS: A gentle introduction*. Medium. https://medium.com/data-reply-it-datatech/dataops-a-gentle-introduction-2cdb1b718029

Deloitte. (n.d.). *Private equity and venture capital annual review*. Retrieved May 10, 2024, from https://www2.deloitte.com/us/en/pages/financial-services/articles/private-equity-venture-capital-annual-review.html

DeMello, M. (2024, June 18). *Artificial intelligence at AstraZeneca*. Emerj. Accessed July 10, 2024. https://emerj.com/ai-sector-overviews/artificial-intelligence-at-astrazeneca

Department of Computer Science and Engineering. (n.d.). *Big data analytics*. National Institute of Technology Silchar. Retrieved May 10, 2024, from *http://cs.nits.ac.in/big-data-analytics/*

Dhade, P., & Shirke, P. (2023). Federated learning for healthcare: A comprehensive review. *Eng. Proc., 59*(1), 230. https://www.mdpi.com/2673-4591/59/1/230

Dhaduk, H. (2022, February 23). *Capital One DevOps case study: A bank with the heart of tech company*. https://www.simform.com/blog/capital-one-devops-case-study/

Dholakia, U. M. (2015, December 21). Everyone hates Uber's surge pricing: Here's how to fix it. *Harvard Business Review*. https://hbr.org/2015/12/everyone-hates-ubers-surge-pricing-heres-how-to-fix-it

Dispenza, J. 2017. *Becoming Supernatural: How common people are doing the uncommon*. Hay House, Inc.

Dobler, Y. (2022, October 4). *Uber knows you: How data optimizes our rides*. Harvard Business School Digital Initiative. https://d3.harvard.edu/platform-digit/submission/uber-knows-you-how-data-optimizes-our-rides/

Dominic, J., & Johnson, K. (2022, March 22). *How the lakehouse democratizes data to help Amgen speed drug development & delivery*. Databricks. https://www.databricks.com/blog/2022/03/22/amgen-modernizes-analytics-with-a-unified-data-lakehouse-to-speed-drug-development-delivery.html

Downs, B. (2023, November 10). *Sutter Health announces ransomware attack that exposed personal information of patients*. CBS News. https://www.cbsnews.com/sacramento/news/sutter-health-announces-ransomware-attack-that-exposed-personal-information-of-patients/

Enderle, R. (2022, January 31). *IBM earnings: Shifting from hardware*. Datamation. https://www.datamation.com/big-data/ibm-earnings-shifting-from-hardware/

Engelbrecht, C. (2017, June 20). *Interactive storytelling on Netflix: Choose what happens next*. Netflix. https://about.netflix.com/en/news/interactive-storytelling-on-netflix-choose-what-happens-next

Engler, A. (2022, June 8). *The EU AI Act will have global impact but a limited Brussels Effect*. Brookings. https://www.brookings.edu/articles/the-eu-ai-act-will-have-global-impact-but-a-limited-brussels-effect/

Ensono. (n.d.). *A case study of DevOps at Netflix*. Retrieved May 10, 2024, from https://www.ensono.com/insights-and-news/expert-opinions/a-case-study-of-dev-ops-at-netflix/

European Parliament. (2024, March 8). *Artificial Intelligence Act: MEPs adopt landmark law*. https://www.europarl.europa.eu/news/en/press-room/20240308IPR19015/artificial-intelligence-act-meps-adopt-landmark-law

Facebook–Cambridge Analytica data scandal. (2024, May 5). In *Wikipedia*. https://en.wikipedia.org/wiki/Facebook%E2%80%93Cambridge_Analytica_data_scandal

Faife, C. (2022, July 6). *The Marriott hotel chain has been hit by another data breach*. The Verge. https://www.theverge.com/2022/7/6/23196805/marriott-hotels-maryland-data-breach-credit-cards

Fellows, B. (2023, August 18). *How to improve the ROI of your software investments*. Loop. https://www.workwithloop.com/blog/how-to-improve-the-roi-of-your-software-investments

Fernández-Manzano, E.-P., Neira, E., & Clares-Gavilán, J. (2016). Data management in audiovisual business: Netflix as a case study. *El Profesional de la Información, 25*(4), 568. https://www.researchgate.net/publication/305741976_Data_management_in_audiovisual_business_Netflix_as_a_case_study

Fidelity. (n.d.). *AI in wealth management*. Retrieved May 10, 2024, from https://institutional.fidelity.com/app/item/RD_9910762/ai-in-wealth-management.html

Finnegan, M. (2024, June 6). Adobe Experience Platform gets AI assistant for customer data insights. *Computerworld*. https://www.computerworld.com/article/2138773/adobe-experience-platform-gets-ai-assistant-for-customer-data-insights.html

Fisher College of Business. (n.d.). *Resilience, focus, and a broader perspective: The case of Airbnb*. Retrieved May 10, 2024, from https://fisher.osu.edu/blogs/leadreadtoday/resilience-focus-and-broader-perspective-case-airbnb

Flinders, K. (2014, January 15). Big banks' legacy IT systems could kill them. *Computer Weekly*. https://www.computerweekly.com/news/2240212567/Big-banks-legacy-IT-systems-could-kill-them

Forrester. (n.d.). *Office Depot: A data-driven execution strategy boosts efficiency and growth*. Retrieved May 10, 2024, from https://www.forrester.com/case-studies/office-depot-a-data-driven-execution-strategy-boosts-efficiency-and-growth/

Frické, M. H. (2018). Data-information-knowledge-wisdom (DIKW) pyramid, framework, continuum. In L. Schintler & C. McNeely (Eds.), *Encyclopedia of big data*. Springer. https://link.springer.com/10.1007/978-3-319-32001-4_331-1

Fruhlinger, J. (2020, February 12). *Equifax data breach FAQ: What happened, who was affected, what was the impact?* CSO. https://www.csoonline.com/article/567833/equifax-data-breach-faq-what-happened-who-was-affected-what-was-the-impact.html

Galea-Pace, S. (2020, May 17). *How Amazon uses big data to transform operations*. Supply Chain Digital. https://supplychaindigital.com/technology/how-amazon-uses-big-data-transform-operations

Gardner, T. (2023, October 4). *How Spotify's UI/UX design helped them win audiences?* Vlinkinfo. https://www.vlinkinfo.com/blog/how-spotifys-ui-ux-design-helped-them-win/

Garfield, S. (2024, March 26). *Adobe and Databricks develop a partnership that enables brands to deliver AI-driven personalization at scale*. Adobe Experience Cloud Blog. https://business.adobe.com/blog/the-latest/adobe-and-databricks-develop-a-partnership-that-enables-brands-to-deliver-ai-driven-personalization-at-scale

Gartner. (2023, March 21). *Gartner survey reveals less than half of data and analytics teams effectively provide value to the organization*. https://www.gartner.com/en/newsroom/press-releases/03-21-2023-gartner-survey-reveals-less-than-half-of-data-and-analytics-teams-effectively-provide-value-to-the-organization

Gelbrich, L. (n.d.). *Going right*. free+style. Retrieved May 10, 2024, from https://freestyleconnection.com/going-right-logan-gelbrich/

Global. (n.d.). *Gaming & interactive solutions: Digital strategy & lead capture optimization, cloud architecture & security*. Retrieved May 10, 2024, from https://globaldataops.com/interactive-gaming/

GlobalData. (2023, December 19). *Amazon.com: Digital transformation strategies*. https://www.globaldata.com/store/report/amazon-com-enterprise-tech-analysis/

Gokul. (n.d.). *Decoding Amazon's recommendation system*. Argoid. Retrieved May 10, 2024, from https://www.argoid.ai/blog/decoding-amazons-recommendation-system

Goldman Sachs. (n.d.). *The embedded finance journey: Innovation that differentiates the customer experience*. Retrieved May 10, 2024, from https://www.goldmansachs.com/what-we-do/transaction-banking/insights/baas.pdf

Google. (2016, July 20). *DeepMind AI reduces energy used for cooling Google data centers by 40%*. Available at: https://blog.google/outreach-initiatives/environment/deepmind-ai-reduces-energy-used-for

Google AI. (n.d.). *Health AI*. Retrieved May 10, 2024, from https://ai.google/discover/healthai/

Google Cloud. (n.d.-a). *Enterprise knowledge graph overview*. Retrieved May 10, 2024, from https://cloud.google.com/enterprise-knowledge-graph/docs/overview

Google Cloud. (n.d.-b). *Optimization AI*. Retrieved May 10, 2024, from https://cloud.google.com/optimization

Google for Developers. (2021, January 28). *How we're helping developers with differential privacy*. https://developers.googleblog.com/2021/01/how-were-helping-developers-with-differential-privacy.html

Google Privacy & Terms. (n.d.). *How Google anonymizes data*. Retrieved May 10, 2024, from https://policies.google.com/technologies/anonymization?hl=en-US

Gudiksen, K. (2019, September 27). *[Case watch] UCFW & Employers Benefit Trust v. Sutter Health: A look at the alleged anticompetitive contract terms in the legal action against Sutter*. The Source on Healthcare Price & Competition. https://sourceonhealthcare.org/case-watch-ucfw-employers-benefit-trust-v-sutter-health-a-look-at-the-alleged-anticompetitive-contract-terms-in-the-legal-action-against-sutter/

Guest, M. (2021, September 29). *Chapter 5: Fujifilm: Innovating out of a crisis*. Medium. https://marcusguest.medium.com/surviving-collapse-in-your-core-market-the-fujifilm-case-3add3159933b

Gupta, S. K. (2023, November 1). *Netflix unveiled: A data-driven case study in streaming strategies*. Medium. https://medium.com/@ffactory335/netflix-unveiled-a-data-driven-case-study-in-streaming-strategies-ac6ac6e56be0

Handa, A., & Vashisht, K. (2017, February 6). *How IBM is embracing the future through design*. UXmatters. https://www.uxmatters.com/mt/archives/2017/02/how-ibm-is-embracing-the-future-through-design.php

Hardy, A.-C. (2016, February 29). *Agile team organisation: Squads, chapters, tribes and guilds*. Medium. https://achardypm.medium.com/agile-team-organisation-squads-chapters-tribes-and-guilds-80932ace0fdc

HBR IdeaCast, Episode 903. (n.d.). *X's Astro Teller on managing moonshot innovation*. Harvard Business Review. Retrieved May 10, 2024, from https://hbr.org/podcast/2023/03/xs-astro-teller-on-managing-moonshot-innovation

Hemachandran, B. (2023, July 13). *Generative AI for efficient test data generation*. LambdaTest. https://www.lambdatest.com/blog/generative-ai-for-efficient-test-data-generation/

Hespell, R. (2023, September 26). *A timeline of Google's biggest AI and ML moments*. Google Blog. https://blog.google/technology/ai/google-ai-ml-timeline/

Hillemann, D. (2023, Janurary 17). *Comparing Open AI and Google DeepMind: Who is leading the AI revolution?* Medium. https://dhillemann.medium.com/comparing-open-ai-and-google-deepmind-who-is-leading-the-ai-revolution-4b11faf07e58

Hiscox. (n.d.). *Adapt or die: Eight businesses transformed their business models to survive*. Retrieved May 10, 2024, from https://www.hiscox.co.uk/broker/about-hiscox/news/adapt-or-die-eight-businesses-transformed-their-business-models-survive

Hitachi Vantara Corporation. (2021, August 19). *Hitachi Vantara delivers intelligent DataOps software suite to give organizations faster access to better data*. PR Newswire. https://www.prnewswire.com/news-releases/hitachi-vantara-delivers-intelligent-dataops-software-suite-to-give-organizations-faster-access-to-better-data-301358821.html

Hitachi Vantara Corporation. (2022, March 16). *Hitachi Vantara unlocks data-driven innovation through an edge-to-cloud data fabric built on its intelligent DataOps portfolio for enterprise and industrial customers*. PR Newswire. https://www.prnewswire.com/news-releases/hitachi-vantara-unlocks-data-driven-innovation-through-an-edge-to-cloud-data-fabric-built-on-its-intelligent-dataops-portfolio-for-enterprise-and-industrial-customers-301503673.html

Hivelr. (2024, July 7). *The economics of Airbnb*. https://www.hivelr.com/2024/07/the-economics-of-airbnb/

Hoag, F. (2012, February 23). *Predictive analytics, informed consent, and privacy: The case of Target*. Foley Hoag. https://www.lexology.com/library/detail.aspx?g=d74d7109-67af-415a-bc18-6c19789dabbf

Human Synergistics. (2024, May 14). *Case study: IBM "reinventing through culture change."* https://www.human-synergistics.com.au/resources/case-study-ibm-reinventing-through-culture-change/

Ibarra, H., & Rattan, A. (2016). *Satya Nadella at Microsoft: Instilling a growth mindset*. London Business School. https://publishing.london.edu/cases/satya-nadella-at-microsoft-instilling-a-growth-mindset/

IBM. (n.d.). *AI ethics*. Retrieved May 10, 2024, from https://www.ibm.com/impact/ai-ethics

IBM Security. (n.d.). *IBM cost of a data breach*. Spirion. Retrieved May 10, 2024, from https://explore.spirion.com/education-data-breaches/ibm-cost-of-a-data-b

Informatica. (n.d.). Gilead Sciences maximizes data value with Informatica. Retrieved May 10, 2024, from https://www.informatica.com/about-us/customers/customer-success-stories/gilead.html

Infinite Monkey. (2022, March). *Embedding Data Science in Cross-Functional Teams.* Available at: https://medium.com/@Infinite_Monkey/embedding-data-science-in-cross-functional-teams-7bfce9283ad2

Inrupt. (n.d.). *The BBC shows its audience the future of personal data access and consent*. Accessed July 10, 2024. https://www.inrupt.com/case-study/bbc-embraces-future-of-personal-data-access-and-consent

Integra IT. (2024, April 30). *How artificial intelligence is transforming clinical trials: The case of AstraZeneca*. Accessed July 10, 2024. https://www.integrait.co/artificial-intelligence-in-clinical-trials-astrazeneca

Intelloz Consulting Group. (2023, May 10). *Discover how Spotify mastered the art of music discoverability: A case study*. LinkedIn. https://www.linkedin.com/pulse/discover-how-spotify-mastered-art-music-discoverability-case

Ishalli. (2023, March 31). *Nokia failure: Learnings from the 4 major causes*. InspireIP. https://inspireip.com/nokia-failure-apple-samsung-or-nokia-itself/

Izrailevsky, Y., Vlaovic, S., & Meshenberg, R. (2016, February 11). *Completing the Netflix cloud migration*. About Netflix. https://about.netflix.com/en/news/completing-the-netflix-cloud-migration

Jerzewski, M. (2023, September 5). *2023 cost of a data breach: Key takeaways*. Fortra. https://www.tripwire.com/state-of-security/cost-data-breach-key-takeaways

Joshi, D., & Spens, J. (2022, October 25). *Data mesh: Real examples and lessons learned*. ThoughtWorks. https://www.thoughtworks.com/en-us/insights/blog/data-engineering/data-mesh-real-examples-and-lessons-learned

Joshi, S. (2020, August 28). *How Warby Parker disrupted the eyewear market*. The Strategy Story. https://thestrategystory.com/2020/08/28/warby-parker-business-model/#google_vignette

JPMorgan Chase (n.d.). *Global financial crimes compliance*. Retrieved May 10, 2024, from https://www.jpmorganchase.com/about/global-financial-crimes-compliance

Karatas, G. (2024, January 2). *Master data management: Best practices & real life examples in '24*. AIMultiple Research. https://research.aimultiple.com/master-data-management/

Kaushik, V. (2024, January 9). *How do Spotify's UI/UX strategies maximize revenue and user engagement?* Techahead. https://www.techaheadcorp.com/blog/how-do-spotifys-ui-ux-strategies-maximize-revenue-and-user-engagement/

Kemp, S. (2021, January 27). *Digital 2021: Global overview report*. DataReportal. https://datareportal.com/reports/digital-2021-global-overview-report

Kent, P. (2012, October 19). *Viewpoint: Big data and big analytics means better business*. BBC News. https://www.bbc.co.uk/news/business-19969588

Kenton, W. (2023, December 18). *Lean Six Sigma: Definition, principles, and benefits*. https://www.investopedia.com/terms/l/lean-six-sigma.asp

Kin + Carta. (2023, August 3). *Building intelligent banking experiences with AI*. https://www.kinandcarta.com/en-us/insights/2023/08/intelligent-banking/

Kinias, Z., & Henderson, F. A. (2020, February 24). *Challenging assumptions about flexible work*. INSEAD Knowledge. https://knowledge.insead.edu/leadership-organisations/challenging-assumptions-about-flexible-work

Klau, R. (2012, October 25). *How Google sets goals: OKRs*. Medium. https://library.gv.com/how-google-sets-goals-okrs-a1f69b0b72c7

Klein, P. (2023, August 30). How an insurance company is attracting the least likely customers: Younger people. *Forbes*. https://www.forbes.com/sites/pklein/2023/08/30/how-an-insurance-company-is-attracting-the-least-likely-customers-younger-people/

KonMari. (n.d.). *What is the KonMari Method™?* Retrieved May 10, 2024, from https://konmari.com/about-the-konmari-method/

Kovalskiy, V. (2024, February 1). *The ROI of AI: Measuring the impact of artificial intelligence investments in business*. Bitrix24.https://www.bitrix24.com/articles/the-roi-of-ai-measuring-the-impact-of-artificial-intelligence-investments-in-business.php

Kraft, A. (2016, March 25). *Microsoft shuts down AI chatbot after it turned into racist Nazi*. CBS News. https://www.cbsnews.com/news/microsoft-shuts-down-ai-chatbot-after-it-turned-into-racist-nazi/

Krebs on Security. (2015, September 21). *Inside Target Corp days after 2013 breach*. https://krebsonsecurity.com/2015/09/inside-target-corp-days-after-2013-breach/

Krishna, H. (2021, July 13). *5 microservices examples: Amazon, Netflix, Uber, Spotify, and Etsy*. SayOne. https://www.sayonetech.com/blog/5-microservices-examples-amazon-netflix-uber-spotify-and-etsy/

Kubernetes. (n.d.). *Case study: Supporting fast decisioning applications with Kubernetes*. Retrieved May 10, 2024, from https://kubernetes.io/case-studies/capital-one/

Kumar, A., & Fioretti, F. (2023, September 28). *The role of artificial intelligence in revolutionising drug discovery*. Eradigm Consulting. Accessed July 10, 2024. https://eradigm.com/white-paper/ai-revolutionizing-drug-discovery

La-Gaffe, G. (2023, October 28). At Stitch Fix, advanced AI is reimagining consumer style. *FAD Magazine*. https://fadmagazine.com/2023/10/28/at-stitch-fix-advanced-ai-is-reimagining-consumer-style/

Lakhiani, V. (2016). *The code of the extraordinary mind: 10 unconventional laws to redefine your life and succeed on your own terms*. Rodale Books.

Lee, D., & Yoon, S. N. (2021). Application of artificial intelligence-based technologies in the healthcare industry: Opportunities and challenges. *Int J Environ Res Public Health* 18(1): 271. https://www.ncbi.nlm.nih.gov/pmc/articles/PMC7795119/

Lee, L. (2021, April 12). *Ethical AI: Where there's a will, there's a way*. The 360 Blog. https://www.salesforce.com/blog/ethical-ai-progress/

Legan, B. (2023, December 7). Maximizing return on innovation investment: 6 ingredients for success. *Supply Chain Management Review*. https://www.scmr.com/article/maximizing-return-on-innovation-investment-6-ingredients-for-success

Li, J., Hudson, S., & So, K. K. F. (2019). Exploring the customer experience with Airbnb. *International Journal of Culture, Tourism and Hospitality Research*. https://www.researchgate.net/publication/335998379_Exploring_the_customer_experience_with_Airbnb

Lill, J. (2023, September 15). *5 ways the power of AI enhances the Adobe Experience Platform*. One North. https://www.onenorth.com/insights/5-ways-the-power-of-ai-enhances-the-adobe-experience-platform/

Lindsey, N. (2019, June 3). Equifax downgrade shows the lasting financial impact of a massive data breach. *CPO Magazine*. https://www.cpomagazine.com/cyber-security/equifax-downgrade-shows-the-lasting-financial-impact-of-a-massive-data-breach/

Loe, M. (2024, February 21). *Oh, Air Canada! Airline pays out after AI accident*. T_HQ. https://techhq.com/2024/02/air-canada-refund-for-customer-who-used-chatbot/

Lopez, J. (2023, July 3). *Analyzing Spotify's UX/UI: A product designer's perspective*. Bootcamp. https://bootcamp.uxdesign.cc/analyzing-spotifys-ux-ui-a-product-designer-s-perspective-4e00eeb3f2ec

Louise, N. (2020, October 8). *IBM to split into two separate companies after 109 years: Plans to focus on cloud computing*. Tech Startups. https://techstartups.com/2020/10/08/ibm-split-two-separate-companies-109-years-plans-focus-cloud-computing/

Ludwig, S., & Fallon-O'Leary, D. (n.d.). *13 hugely successful companies that reinvented their business*. U.S. Chamber of Commerce. Retrieved May 10, 2024, from https://www.uschamber.com/co/good-company/growth-studio/successful-companies-that-reinvented-their-business

Luxton, D. D. (2019, February). Should Watson be consulted for a second opinion? *AMA Journal of Ethics*. https://journalofethics.ama-assn.org/article/should-watson-be-consulted-second-opinion/2019-02

Macht, D. (2023, November 10). *Sutter Health vendor data breach exposes personal information of more than 845,000 patients*. KCRA 3 News. https://www.kcra.com/article/sutter-health-vendor-data-breach-moveit/45807041

Magsino, M. (n.d.). *Putting the fast in fashion: How Zara drastically reduced lead times*. Fohlio. Retrieved May 10, 2024, from https://www.fohlio.com/blog/zaras-fashion-revolution-how-strategic-optimization-of-lead-time-propelled-them-to-the-forefront-of-fast-fashion

Market.us. (2024, January 11). *Federated learning market set to hit USD 311.4 million by 2032*. LinkedIn. https://www.linkedin.com/pulse/federated-learning-market-set-hit-usd-3114-million-2032-markets-us-ubvoc/

MarketsandMarkets. (n.d.). *Artificial intelligence market worth $1,345.2 billion by 2030*. Retrieved May 10, 2024, from https://www.marketsandmarkets.com/PressReleases/artificial-intelligence.asp

Marr, B. (2018, October 5). How does Amazon's Alexa really work? *Forbes*. https://www.forbes.com/sites/bernardmarr/2018/10/05/how-does-amazons-alexa-really-work/?sh=5fb7078d1937

Marr, B. (2023, April 28). The amazing ways Duolingo is using AI and GPT-4. *Forbes*. https://www.forbes.com/sites/bernardmarr/2023/04/28/the-amazing-ways-duolingo-is-using-ai-and-gpt-4/?sh=5bd3794f1346

Maskalevich, T. (n.d.). Stitch Fix unravels AI behind personalized styling. Interactions Podcast (The Conversaition: Season 4 Episode 24). Retrieved July 30, 2024, from https://www.interactions.com/podcasts/stitch-fix-unravels-ai-behind-personalized-styling/

Mauro, T. (2015, February 19). *Adopting microservices at Netflix: Lessons for architectural design*. F5. https://www.f5.com/company/blog/nginx/microservices-at-netflix-architectural-best-practices

MBA Knowledge Base. (n.d.). *Case study: Inventory management practices at Walmart*. Retrieved May 10, 2024, from https://www.mbaknol.com/management-case-studies/case-study-of-walmart-inventory-management/

McKinsey & Company. (2021, December 7). *Losing from day one: Why even successful transformations fall short*. https://www.mckinsey.com/capabilities/people-and-organizational-performance/our-insights/successful-transformations

McKinsey & Company. (n.d.). *The rise of quantum computing*. Retrieved May 10, 2024, from https://www.mckinsey.com/featured-insights/the-rise-of-quantum-computing

Meier, K., & Spichiger, R. (2024, March 15). *The EU AI regulation is coming: What does it mean for you and your business in Switzerland?* EY. https://www.ey.com/en_ch/forensic-integrity-services/the-eu-ai-act-what-it-means-for-your-business

Menczer, F. (2021, September 21). *Facebook algorithm*. The Fulcrum. https://thefulcrum.us/media-technology/facebook-algorithm

Meredith, S. (2018, April 10). *Facebook-Cambridge Analytica: A timeline of the data hijacking scandal*. CNBC. https://www.cnbc.com/2018/04/10/facebook-cambridge-analytica-a-timeline-of-the-data-hijacking-scandal.html

Meta. (n.d.). Meta careers: Connecting the world through AI and other emerging technologies. Accessed July 10, 2024, from https://www.metacareers.com

Meta. (2022, June). *Research highlights from the Core Data Science team at Meta.* Available at: https://research.facebook.com/blog/2022/6/research-highlights-from-the-core-data-science-team-at-meta

Metz, C. (2023, August 29). The A.I. revolution is coming: But not as fast as some people think. *New York Times*. https://www.nytimes.com/2023/08/29/technology/ai-revolution-time.html

Mihaila, T., & Janson, E. (2021, September 23). *When experience is not enough: Using data and analytics throughout the PE investment cycle*. PwC. https://www.pwc.com/us/en/tech-effect/ai-analytics/applying-data-and-analytics-in-private-equity-firms.html

ModelOp. (n.d.). *AI model operations at scale for Fidelity Investments*. Retrieved May 10, 2024, from https://www.modelop.com/blog/ai-model-operations-at-scale-for-fidelity-investments

Mondal, B. (2020, September 16). *Netflix's big data architecture*. Medium. https://medium.com/@bishal135/netflixs-big-data-architecture-c27d90ba781e

Mondal, D. (2023, July 19). *Saxo Bank improves data governance with data mesh*. LinkedIn. https://www.linkedin.com/pulse/saxo-bank-improves-data-governance-mesh-debajit-mondal

Morein, G. (n.d.). *7 out of 7: Success during COVID-19*. ThinkLouder. Retrieved May 10, 2024, from https://thinklouder.com/companies-successfully-adapted-covid19/

Mpeshev. (2023, May 16). *The power of data: How Amazon utilizes big data to drive sales*. CEO Hangout. https://ceohangout.com/the-power-of-data-how-amazon-utilizes-big-data-to-drive-sales/

Mulkers, Y. (2023, May 18). *Exploring how Spotify uses data analytics effectively*. 7wData. https://7wdata.be/analytics/how-spotify-uses-data-analytics/

Murphy. (2023, March 13). *The impact of Spotify on the music industry*. SocialBoosting. https://www.socialboosting.com/blog/spotify/the-impact-of-spotify-on-the-music-industry

Naess, K. (2021, March 17). *Case study in DataOps: How data puts the game in gaming*. Hitachi Vantara. https://www.hitachivantara.com/en-us/company/events-and-webinars/case-study-in-dataops-how-data-puts-the-game-in-gaming

Najibi, A. (2020, October 24). *Racial discrimination in face recognition technology*. Harvard University: Science in the News. https://sitn.hms.harvard.edu/flash/2020/racial-discrimination-in-face-recognition-technology/

Nasser, M. J. (2023, September 24). *Revolutionizing hospitality: Airbnb's success story powered by B2B and SaaS innovation*. LinkedIn. https://www.linkedin.com/pulse/revolutionizing-hospitality-airbnbs-success-story-powered-nasser

Natalie. (2017, April 5). *Netflix: A personalized viewing experience*. Harvard Business School Digital Initiative. https://d3.harvard.edu/platform-digit/submission/netflix-a-personalized-viewing-experience/

Nehme, A. (2023, April). *Overcoming top challenges in data upskilling: Insights from the State of Data Literacy 2023 Report*. DataCamp. https://www.datacamp.com/blog/overcoming-top-challenges-in-data-upskilling

Nemire, B. (2016, September 15). *Microsoft's voice recognition technology almost as accurate as humans*. NVIDIA Developer. https://developer.nvidia.com/blog/microsofts-voice-recognition-technology-almost-as-accurate-as-humans/

Netflix. (n.d.). *How Netflix's recommendations system works*. Accessed May 27, 2024. https://help.netflix.com/en/node/100639

Netflix Technology Blog. (2013, March 27). *System architectures for personalization and recommendation*. Medium. https://netflixtechblog.com/system-architectures-for-personalization-and-recommendation-e081aa94b5d8

Nitro Logistics. (2023, March 16). *Shopify success stories: Inspiring case studies of thriving e-commerce businesses*. https://nitrologistics.com/blog/shopify-success-stories-inspiring-case-studies-of-thriving-e-commerce

O'Brien, C. (2023, June 14). *The unstoppable success of Netflix*. Digital Marketing Institute. https://digitalmarketinginstitute.com/blog/the-unstoppable-success-of-netflix

OpenAI. (n.d.). *OpenAI Charter*. Retrieved May 10, 2024, from https://openai.com/charter

Ostili, L. (2024, April 1). *The best way to maximize ROI in business development*. LinkedIn. https://www.linkedin.com/pulse/best-way-maximize-roi-business-development-lorenzo-ostili-jtpwf

Owen, R. (2021, October 4). *Artificial intelligence at Airbnb: Two unique use-cases*. Emerj. https://emerj.com/ai-sector-overviews/artificial-intelligence-at-airbnb/

Owen, R. (2022, January 24). *Artificial intelligence at PayPal: Two unique use-cases*. Emerj. https://emerj.com/ai-sector-overviews/artificial-intelligence-at-paypal/

Palumbo, D. (2021, February 7). *Amazon: The unstoppable rise of the internet giant*. BBC News. https://www.bbc.com/news/business-55927979

Pappas, C. (2023, December 27). *6 strategies for maximizing ROI on employee training investments*. eLearning Industry. https://elearningindustry.com/strategies-for-maximizing-roi-on-employee-training-investments

Parati. (2023, December 22). *How AI enhances OKR management: Aligning intelligence with goals*. LinkedIn. https://www.linkedin.com/pulse/how-ai-enhances-okr-management-aligning-intelligence-goals-ebx6c/

Pasick, A. (2015, December 21). *The magic that makes Spotify's Discover Weekly playlists so damn good*. Quartz. https://qz.com/571007/the-magic-that-makes-spotifys-discover-weekly-playlists-so-damn-good

Patel, B. N., et al. (2019). Human–machine partnership with artificial intelligence for chest radiograph diagnosis. *npj Digit. Med. 2*(111). https://www.nature.com/articles/s41746-019-0189-7

Patel, S., & Shetty, S. (2024, June 11). *Advancing personal health and wellness insights with AI*. Google Research Blog. https://research.google/blog/advancing-personal-health-and-wellness-insights-with-ai/

Paul, S. (2020, October 7). *Case study: How Netflix uses AWS for innovation, agility and scalability*. LinkedIn. https://www.linkedin.com/pulse/case-study-how-netflix-uses-aws-innovation-agility-scalability-paul

Peralta, R. (2002, June 23). *Alan Turing's everlasting contributions to computing, AI, and cryptography*. National Institute of Standards and Technology (NIST). https://www.nist.gov/blogs/taking-measure/alan-turings-everlasting-contributions-computing-ai-and-cryptography

Peranandam, C. (2018, December). AI helps Duolingo personalize language learning. *Wired*. https://www.wired.com/brandlab/2018/12/ai-helps-duolingo-personalize-language-learning/

Pickup, E. (2021, September 7). The Oversight Board's dormant power to review Facebook's algorithms. *Yale Journal on Regulation.* https://www.yalejreg.com/bulletin/the-oversight-boards-dormant-power-to-review-facebooks-algorithms/

Porter, J. (2019, October 23). Google confirms "quantum supremacy" breakthrough. *The Verge*. https://www.theverge.com/2019/10/23/20928294/google-quantum-supremacy-sycamore-computer-qubit-milestone

Powderly, H. (2015, September 16). *Sutter Health says data on 2,500 patients involved in potential breach*. Healthcare Finance. https://www.healthcarefinancenews.com/news/sutter-health-says-data-2500-patients-involved-potential-breach

Pragmatic Institute. (n.d.). *Case study: How Spotify prioritizes data projects for a personalized music experience*. Retrieved May 10, 2024, from https://www.pragmaticinstitute.com/resources/articles/data/case-study-how-spotify-prioritizes-data-projects-for-a-personalized-music-experience/

Press, G. (2021, December 24). On becoming data-driven: Today's business imperative. *Forbes*. https://www.forbes.com/sites/gilpress/2021/12/24/on-becoming-data-driven-todays-business-imperative/

Process.st. (n.d.). *How to automate Fidelity Investments*. Retrieved May 10, 2024, from https://www.process.st/how-to/automate-fidelity-investments

ProjectPro. (2024, April 14). *How data science increased Airbnb's valuation to $25.5 bn**. https://www.projectpro.io/article/how-data-science-increased-airbnbs-valuation-to-25-5-bn/199

Pulitzer Prizes. (n.d.). *The Guardian US*. Accessed July 10, 2024. https://www.pulitzer.org/winners/guardian-us

Q.ai. (2022, September 29). Tesla: A history of innovation (and headaches). *Forbes*. https://www.forbes.com/sites/qai/2022/09/29/tesla-a-history-of-innovation-and-headaches/

quade. (2024, April 1). *Adobe unveils their new Adobe Experience Platform AI assistant*. GeniusOS. Accessed June 14, 2024. https://geniusos.co/adobe-unveils-their-new-adobe-experience-platform-ai/

QuantumBlack: AI by McKinsey. (2022, January 28). *The data-driven enterprise of 2025*. https://www.mckinsey.com/capabilities/quantumblack/our-insights/the-data-driven-enterprise-of-2025

Qumer, S. M., & Ikrama, S. (2021). *Zoom's rise amidst the COVID-19 pandemic*. IBS Center for Management Research. https://www.thecasecentre.org/products/view?id=175471

R1. (2022, July 14). *R1 RCM announces 10-year end-to-end RCM partnership with Sutter Health*. https://www.r1rcm.com/news-and-press/r1-rcm-announces-10-year-end-to-end-rcm-partnership-with-sutter-health/

Rakhra, C. (2023, April 12). *Case study: 2 Netflix*. Medium. https://medium.com/@chetxn/case-study-2-netflix-3e1ccb70d206

Ranadive, A. (2016, January 11). *Lessons from Pixar 1: The braintrust*. Medium. https://medium.com/great-business-stories/lessons-from-pixar-1-the-braintrust-e306843a5153

Ranosys. (2021, August 3). *Delivering highly personalized experiences with Adobe Sensei*.

https://www.ranosys.com/blog/insights/personalized-experiences-with-adobe-sensei/

Ransbotham, S. (2024, April 16). *Fashioning the perfect fit with AI: Stitch Fix's Jeff Cooper*. MIT Sloan Management Review Podcast (Me, Myself, and AI: Episode 903). https://sloanreview.mit.edu/audio/fashioning-the-perfect-fit-with-ai-stitch-fixs-jeff-cooper/

Rassweiler, A., & Brinley, S. (2014, October 7). *Tesla Motors: A case study in disruptive innovation*. S&P Global Mobility. https://www.spglobal.com/mobility/en/research-analysis/q14-tesla-motors-a-case-study-in-disruptive-innovation.html

Ravulakollu, N. (2023, May 16). *Netflix churn prediction case study*. AlmaBetter. https://www.almabetter.com/bytes/articles/netflix-churn-prediction-case-study

Reqtest. (2015, March 30). *How Spotify does agile: A look at the Spotify engineering culture*. https://reqtest.com/en/knowledgebase/how-spotify-does-agile-a-look-at-the-spotify-engineering-culture/

Revelate. (2023, October 3). *Data mesh principles: A new approach data management*. https://revelate.co/blog/data-mesh-principles-a-new-approach-data-management/

Ribeiro, R. (2015, January 26). Google X co-founder Tom Chi says innovation comes from doing, not guessing. *BizTech Magazine*. https://biztechmagazine.com/article/2015/01/google-x-co-founder-tom-chi-says-innovation-comes-doing-not-guessing

Robert H. Smith School of Business at the University of Maryland. (n.d.). *What happened to Yahoo in 6 points*. Retrieved May 10, 2024, from https://onlinebusiness.umd.edu/blog/what-happened-to-yahoo-in-6-points/

Robertson, G. (n.d.). *Lego case study: How to revitalize a beloved brand*. Beloved Brands. Retrieved May 10, 2024, from https://beloved-brands.com/lego-case-study/

Rogers, S. (2021, September 13). Numbers you can tell stories with: A decade of *Guardian* data journalism. *Guardian*. https://www.theguardian.com/membership/datablog/2021/sep/13/numbers-you-can-tell-stories-with-a-decade-of-guardian-data-journalism

Rohn, S. (2022, July 7). *6 high-profile digital transformation failures (+causes)*. Whatfix. https://whatfix.com/blog/digital-transformation-failures/

Rosser, J. (n.d.). *Spotify: A case study in business strategy and value compounding*. MOI Global. Retrieved May 10, 2024, from https://moiglobal.com/spotify-case-study-202008/

Sachdeva, A. (2023, April 18). *5 data-driven decision-making examples*. GapScout. https://gapscout.com/blog/5-data-driven-decision-making-examples/

Sadeh, G. (2019, March 20). *How Netflix uses big data content*. ClickZ. https://www.clickz.com/how-netflix-uses-big-data-content/

Saradhi V., P. (2023, September 29). *Real-world success stories: Transforming data platforms by addressing underlying issues*. LinkedIn. https://www.linkedin.com/pulse/real-world-success-stories-transforming-data-platforms-v

Saripalli, S. (2024, March 28). *What is DataOps?* Sprint2Scale. https://sprint2scale.com/what-is-dataops/

Satell, G. (2014, September 5). A look back at why Blockbuster really failed and why it didn't have to. *Forbes*. https://www.forbes.com/sites/gregsatell/2014/09/05/a-look-back-at-why-blockbuster-really-failed-and-why-it-didnt-have-to/?sh=33c1fe111d64

Saunders, A., Vaidya, N., Tetelman, A., & Spirin, N. (2024, March 18). *NVIDIA NIM offers optimized inference microservices for deploying AI models at scale*. NVIDIA Developer. https://developer.nvidia.com/blog/nvidia-nim-offers-optimized-inference-microservices-for-deploying-ai-models-at-scale/

Sawhney, M., Shah, B., Yu, R., Rubtsov, E., & Goodman, P. (2020). *Uber: Applying machine learning to improve the customer pickup experience*. Kellogg School of Management. https://www.emerald.com/insight/content/doi/10.1108/case.kellogg.2021.000090/full/html

Schneider, A. (2018, November 30). *Marriott says up to 500 million customers' data stolen in breach*. NPR. https://www.npr.org/2018/11/30/672167870/marriott-says-up-to-500-million-customers-data-stolen-in-breach

Shah, A. (2023, March 8). *How Tesla uses and improves its AI for autonomous driving*. EnterpriseAI. https://www.enterpriseai.news/2023/03/08/how-tesla-uses-and-improves-its-ai-for-autonomous-driving/

Shahid, K. (2023, September 26). *A look back at Google's biggest artificial intelligence projects*. HubSpot Blog. https://blog.hubspot.com/marketing/a-look-back-at-googles-biggest-artificial-intelligence-projects

Shakir, T. (2022, September 27). *DataOps: A practical understanding and implementation.* LinkedIn. https://www.linkedin.com/pulse/dataops-practical-understanding-implementation-taaha-khan-

Sharpen. (n.d.). *How Netflix moved operations to the cloud and saw revenue boom: A digital transformation case* study. Retrieved May 10, 2024, from https://sharpencx.com/blog/netflix-digital-transformation-case-study/

Sheeran, D. (2024, March 6). *Key takeaways from Gilead's Innovation Talk at re:Invent 2023*. Amazon Web Services. https://aws.amazon.com/blogs/industries

Shopsys. (n.d.). *Case study: Office Depot*. Retrieved May 10, 2024, from https://www.shopsys.com/office-depot-case-study/

Shrivastava, D. & Hawelia, A. (2024, January 9). *9 reasons why Nokia failed after enjoying unrivaled dominance*. StartupTalky. https://startuptalky.com/reasons-why-nokia-failed/

Sims, J., Ameen, N., & Bauer, R. (2019). Dynamic pricing and benchmarking in Airbnb. *UK Academy for Information Systems Conference Proceedings 2019*, 36. https://aisel.aisnet.org/ukais2019/36

Sinek, S. (2011). *Start with why: How great leaders inspire everyone to take action*. Portfolio.

Sinha, A. (2023, November 17). *Emerging practices for society-centered AI*. Google Research. https://research.google/blog/emerging-practices-for-society-centered-ai/

Sivarajah, U., Kamal, M. M., Irani, Z., & Weerakkody, V. (2017). Analyzing the impact of digital marketing on consumer behavior. *Journal of Business Research*, 79(4), 435–448. https://www.sciencedirect.com/science/article/pii/S014829631630488X

Slaveykova, K. (2017, December 9). *How Netflix uses data to keep you binge-watching & personalize your viewing experience*. Medium. https://medium.com/the-data-nudge/how-netflix-uses-data-to-keep-you-binge-watching-personalize-your-viewing-experience-894c99a1e2b4

SmartBear. (n.d.). *Intuit*. Retrieved May 10, 2024, from https://smartbear.com/resources/case-studies/intuit/

Sobel, S. (2024, March 26). *Delivering next generation consumer experiences: Databricks and Adobe announce strategic partnership*. Databricks. https://www.databricks.com/blog/delivering-next-generation-consumer-experiences-databricks-and-adobe-announce-strategic

Sobers, R. (2023, November 10). *84 must-know data breach statistics [2023]*. Varonis. https://www.varonis.com/blog/data-breach-statistics

Spatharou, A., Hieronimus, S., & Jenkins, J. (2020, March 10). *Transforming healthcare with AI: The impact on the workforce and organizations*. McKinsey & Company. https://www.mckinsey.com/industries/healthcare/our-insights/transforming-healthcare-with-ai

Sperber, L. (2016, February 25). *Building software at Etsy*. Lauren Sperber. https://laurensperber.com/2016/02/25/presentation-building-software-at-etsy/

Stackpole, B. (2023). *The analytics edge: How to turn data into competitive advantage*. MIT Management Sloan School. https://mitsloan.mit.edu/sites/default/files/2023-02/MITSloan-TheAnalyticsEdge.pdf

Steinlage, A. J. (2016, November 17). *Fujifilm: Outlasting the "Kodak moment."* Harvard Business School Digital Initiative. https://d3.harvard.edu/platform-rctom/submission/fujifilm-outlasting-the-kodak-moment/

Stitch Fix. (n.d.). *Algorithms tour*. Retrieved July 30, 2024, from https://algorithms-tour.stitchfix.com/

Stobierski, T. (2019, August 26). *The advantages of data-driven decision-making*. Harvard Business School Online. https://online.hbs.edu/blog/post/data-driven-decision-making

Strutner, S. (2020, May 20). *12 examples of smart business adaptations amid COVID-19*. NetSuite. https://www.netsuite.com/portal/resource/articles/business-strategy/12-examples-of-smart-business-adaptations-amid-covid-19.shtml

Summers, E. (2024, March 28). *Optimize your strategy: A deep dive into Airbnb business overview*. Efinancialmodels. https://www.efinancialmodels.com/optimize-your-strategy-a-deep-dive-into-airbnb-business-overview

Supply Chain Today. (n.d.). *Amazon uses AI to deliver orders even faster*. Retrieved May 10, 2024, from https://www.supplychaintoday.com/amazon-uses-ai-to-deliver-orders-even-faster/

Sutter Health. (n.d.) *Being not-for-profit*. Retrieved May 10, 2024, from https://www.sutterhealth.org/about/being-not-for-profit

Swann, C., & Kelly, M. (2023). Disruption in the meat industry: New technologies in nonmeat substitutes. *Business Economics 58*(1): 42–60, https://doi.org/10.1057/s11369-023-00302-w

Taboola. (2023, January 17). *Yahoo and Taboola announce closing of deal, 30-year strategic partnership sees Taboola power recommendations for Yahoo*. https://www.taboola.com/press-release/taboola-and-yahoo-close-deal

Tacken, D. (2019, May 31). *Office Depot: Valued for distress, but the business is turning around*. Seeking Alpha. https://seekingalpha.com/article/4267608-office-depot-valued-for-distress-business-is-turning-around

Talend. (n.d.). *Office Depot | Viking*. Retrieved May 10, 2024, from https://www.talend.com/customers/office-depot-viking/

TEDx Talks. (2014, February 6). Knowing is the enemy of learning: Tom Chi at TEDxSemesteratSea [Video]. YouTube. https://www.youtube.com/watch?v=_WtsMrkfG1w

Texta. (n.d.). Revolutionizing marketing: Unleashing the power of AI in Unilever's marketing strategy. Texta Blog. Retrieved July 30, 2024, from https://texta.ai/blog-articles/revolutionizing-marketing-unleashing-the-power-of-ai-in-unilevers-marketing-strategy

The Guardian. (2016, July 20). *Google's AI Reduces Energy Consumption in Data Centers by 40%.*

Available at: https://www.theguardian.com/environment/2016/jul/20/google-ai-cut-data-centre-energy-use-15-per-cent

Thinklogic. (n.d.). *Cloud migration case study: How Capital One successfully transitioned to the cloud.* Retrieved May 10, 2024, from https://www.thinklogic.com/post/cloud-migration-case-study-how-capital-one-successfully-transitioned-to-the-cloud

This is Design Thinking. (n.d.). *IBM: Design thinking adaptation and adoption at scale.* Retrieved May 10, 2024, from https://thisisdesignthinking.net/2019/07/ibm-design-thinking-adaptation-adoption-at-scale/

Thompson, N. C., Ge, S., & Manso, G. F. (2022, June 28). The importance of (exponentially more) computing power. *arXiv:2206.14007.* https://arxiv.org/abs/2206.14007

Torrella, K. (2023, April 17). Were the impossible and beyond burgers a fad, or is plant-based meat here to stay? *Vox.* https://www.vox.com/future-perfect/2023/4/17/23682232/impossible-beyond-plant-based-meat-sales

Turon, M. (2023, April 26). *Netflix's secret sauce: How data and analytics propelled them to global domination.* LinkedIn. https://www.linkedin.com/pulse/netflixs-secret-sauce-how-data-analytics-propelled-them-michael-turon

Tyko, K. (2020, April 17). Coronavirus demand: Whole Foods converts some stores to online only. *USA Today.* https://www.usatoday.com/story/money/food/2020/04/17/coronavirus-demand-whole-foods-converts-some-stores-online-only/5141287002/

Uberoi, R. (2017, April 5). *Zara: Achieving the "fast" in fast fashion through analytics.* Digital Initiative at Harvard Business School. https://d3.harvard.edu/platform-digit/submission/zara-achieving-the-fast-in-fast-fashion-through-analytics/

Uchitani, A. (n.d.). *Shu Ha Ri: The Japanese art of mastery.* Azumi Uchitani. Retrieved May 10, 2024, from https://www.azumiuchitani.com/shuhari/

Unilever. (2024a, April 4). *How AI and digital help us innovate faster and smarter.* https://www.unilever.com/news/news-search/2023/how-ai-and-digital-help-us-innovate-faster-and-smarter/

Unilever. (2024b, June 14). *How AI-powered ultra-personalised experiences are boosting our beauty brands.* https://www.unilever.com/news/news-search/2024/how-aipowered-ultrapersonalised-experiences-are-boosting-our-beauty-brands/

Unscrambl. (2021, June 15). *6 inspiring examples of data-driven companies (key takeaways included).* https://unscrambl.com/blog/data-driven-companies-examples/

UPS. (2020, January 29). *UPS to enhance Orion with continuous delivery route optimization.* GlobeNewswire. https://www.globenewswire.com/news-release/2020/01/29/1977072/0/en/UPS-To-Enhance-Orion-With-Continuous-Delivery-Route-Optimization.html

US Department of Justice. (2021, August 30). *Sutter Health and affiliates to pay $90 million to settle false claims act allegations of mischarging the Medicare Advantage program.* https://www.justice.gov/opa/pr/sutter-health-and-affiliates-pay-90-million-settle-false-claims-act-allegations-mischarging

UserTesting. (n.d.). *IBM + UserTesting.* Retrieved May 10, 2024, from https://www.usertesting.com/resources/customers/ibm

van Es, K. (2022). Netflix & big data: The strategic ambivalence of an entertainment company. *Television & New Media, 24*(5). https://www.researchgate.net/publication/363867276_Netflix_Big_Data_The_Strategic_Ambivalence_of_an_Entertainment_Company

Van Kuiken, S. (2022, October 21). *Tech at the edge: Trends reshaping the future of IT and business*. McKinsey & Company. https://www.mckinsey.com/capabilities/mckinsey-digital/our-insights/tech-at-the-edge-trends-reshaping-the-future-of-it-and-business

Vardhman, R. (2024, February 6). *What is the average time spent on TikTok*. DataProt. https://dataprot.net/statistics/average-time-spent-on-tiktok/

Varshneya, K. (2021, December 2021). *Understanding design of microservices architecture at Netflix*. TechAhead. https://www.techaheadcorp.com/blog/design-of-microservices-architecture-at-netflix/

Vashishta, V. (2023). *Vin Vashishta's post*. LinkedIn. https://www.linkedin.com/posts/vineetvashishta_datascience-analytics-hiringprocess-activity-7019669177242185728-3wMi

Velazco, E. (n.d.). *The mistakes that killed Friendster*. Geeks. Retrieved May 10, 2024, from https://vocal.media/geeks/the-mistakes-that-killed-friendster

Venkataraman, S., & Cross, T. (2009). *Office Depot e-business*. Darden Business Publishing. https://www.researchgate.net/publication/228179863_Office_Depot_E-Business

Verint. (n.d.). *Saxo Bank drives compliant, collaborative customer engagement with Verint*. Retrieved May 10, 2024, from https://www.verint.com/case-studies/saxo-bank-drives-compliant-collaborative-customer-engagement-with-verint/

Viégas, F., Holbrook, J., & Wattenberg, M. (2018, December 21). *How we worked to make AI for everyone in 2018*. Google Blog. https://blog.google/technology/ai/how-we-worked-make-ai-everyone-2018/

Vincent, J. (2018, October 10). Amazon reportedly scraps internal AI recruiting tool that was biased against women. *The Verge*. https://www.theverge.com/2018/10/10/17958784/ai-recruiting-tool-bias-amazon-report

Walker, D. (2023, October 31). *At Fidelity, AI provides extra eyes to spot red flags*. Chief Investment Officer. https://www.ai-cio.com/news/at-fidelity-ai-provides-extra-eyes-to-spot-red-flags

Walton, R. (2016, July 21). *Google uses artificial intelligence to boost data center efficiency*. Utility Dive. https://www.utilitydive.com/news/google-uses-artificial-intelligence-to-boost-data-center-efficiency/423086/

Wang, S. (2022). Explanations to the failure of Nokia phone. *2022 7th International Conference on Financial Innovation and Economic Development (ICFIED 2022)*. https://www.researchgate.net/publication/363357979_Explanations_to_the_Failure_of_Nokia_Phone

Watkins, M. D. (2020, January). *10 reasons why organizational change fails*. IMD. https://www.imd.org/research-knowledge/transition/articles/10-reasons-why-organizational-change-fails/

Wawira, J. (2019, June 26). Logistics challenges facing eCommerce in Africa. *Africa Business*. https://africabusiness.com/2019/06/26/logistics-challenges-facing-ecommerce-in-africa/

WBR Insights. (n.d.). *How Stitch Fix uses data science and machine learning to deliver personalization at scale*. eTail. Retrieved July 30, 2024, from https://etailwest.wbresearch.com/blog/how-stitch-fix-uses-data-science-and-machine-learning-to-deliver-personalization-at-scale

Weatherbed, J. (2023, March 23). Canva's got a massive update that should have Adobe worried. *The Verge.* https://www.theverge.com/2023/3/23/23652131/canva-ai-update-visual-worksuite-brand-tools-features

Webb, D. (2020, January 10). *Audience choices: Personalised media*. BBC Research & Development. https://www.bbc.co.uk/rd/blog/2020-01-audience-choices-data-personalised-media

Webwinkelvakdag. (2019, September 26). *DataOps: The next big wave on your data journey*. SlideShare. https://www.slideshare.net/slideshow/dataops-the-next-big-wave-on-your-data-journey-big-data-expo/176309422

Weldon, D. (2024, January 24). *16 top data governance tools to know about in 2024*. TechTarget. https://www.techtarget.com/searchdatamanagement/feature/15-top-data-governance-tools-to-know-about

Wells, J. (2020, September 1). *Whole Foods moves beyond stores for online fulfillment*. Grocery Dive. https://www.grocerydive.com/news/whole-foods-moves-beyond-stores-for-online-fulfillment/584536/

Whitehorn. (n.d.). *Case study: The Shopify surge*. Retrieved May 10, 2024, from https://www.whitehorncapital.com/whitehorn-blog/2020/6/8/case-study-the-shopify-surge

Wired. (2016, July 20). *How Google Used DeepMind AI to Slash Its Data Center Energy Use.* Available at: https://www.wired.com/story/google-deepmind-data-centres-efficiency

Wolverton, T. (2011, July 13). 2011: Backlash grows against Netflix price hike. *Mercury News.* https://www.mercurynews.com/2011/07/13/2011-backlash-grows-against-netflix-price-hike/

World Economic Forum. (2021, April 28). *These are 2021's most innovative companies*. https://www.weforum.org/agenda/2021/04/worlds-most-innovative-companies/

Wosepka, K. (n.d.). *Fintechs make inroads with customers as legacy banks push back*. MIT Management Sloan School. Retrieved May 10, 2024, from https://mitsloan.mit.edu/ideas-made-to-matter/fintechs-make-inroads-customers-legacy-banks-push-back

Yip, P. (2017, January 16). *How to grow a business in 190 Markets: 4 lessons from Airbnb*. OneSky. https://www.oneskyapp.com/blog/airbnb-global-growth/

Young, K. (2021, September 13). *Cyber case study: Target data breach*. CoverLink Insurance. https://coverlink.com/cyber-liability-insurance/target-data-breach/

Zeenea. (2022, October 19). *The guide to becoming data-driven by Airbnb*. https://zeenea.com/the-guide-to-becoming-data-driven-by-airbnb/

Zeleny, M. (2005). *Human systems management: integrating knowledge, management and systems*. World Scientific Pub.

Zheng, H. (2023, December 29). *Gender bias in hiring algorithms*. Medium. https://medium.com/@hzhe0050/gender-bias-in-hiring-algorithms-8a39ea49a594

Zinsmeister, S. (2023, May 8). *What is a data mesh and how does it transform your data architecture?* ThoughtSpot. https://www.thoughtspot.com/data-trends/data-integration/data-mesh

www.ingramcontent.com/pod-product-compliance
Lightning Source LLC
Chambersburg PA
CBHW060926210326
41597CB00042B/4524